SCIENCE AND REALITY

Ernan McMullin

Science and Reality:

Recent Work in the Philosophy of Science

Essays in Honor of Ernan McMullin

EDITED BY

James T. Cushing
C. F. Delaney
Gary M. Gutting

UNIVERSITY OF NOTRE DAME PRESS
NOTRE DAME, INDIANA 46556

Library of Congress Cataloging in Publication Data

Main entry under title:

Science and reality.

  Bibliography: p.
  1. Science — Philosophy — Addresses, essays, lectures.
2. Physics — Philosophy — Addresses, essays, lectures.
3. McMullin, Ernan, 1924– —Addresses, essays,
lectures. I. McMullin, Ernan, 1924– .
II. Cushing, James T., 1937– . III. Delaney, C. F.
(Cornelius F.), 1938– . IV. Gutting, Gary.
Q175.3.S3 1984   501   84-40360
ISBN 0-268-01714-X
ISBN 0-268-01715-8 (pbk.)

57,652

DEDICATED TO ERNAN McMULLIN

Priest, scholar, and educator, Ernan McMullin has been one of the principal voices in the philosophy of science for the last thirty years. While avoiding the extremes of historicism and logicism, he has made major contributions to both the historical and critical traditions in the philosophy of science. His books on Galileo and Newton have become standards, and his studies of explanation and realism have been at the forefront of philosophical discussion.

His contribution to philosophy, moreover, has been more than merely scholarly. He has served his university, Notre Dame, as chairman of its philosophy department for seven years and has served the philosophical community at large to an unprecedented degree. Ernan has been president of the American Philosophical Association, the Philosophy of Science Association, the American Catholic Philosophical Association, and the Metaphysical Society of America, and has served on numerous national and international committees. But in keeping with his principal calling, his most profound impact may well be personal; he is a friend and confidant of uncountably many.

v

Contents

Preface

Many people have generously contributed to making a reality of this volume of essays honoring Ernan McMullin on the occasion of his sixtieth birthday. During the early planning stages and in the final production of the text (especially in accepting responsibility for proofreading the galleys), several of Ernan's colleagues here at Notre Dame played an essential role: Michael J. Crowe, Vaughn R. McKim, A. Edward Manier, and Phillip R. Sloan. Desmond M. Clarke, who was on leave at Notre Dame during the 1982–83 academic year, also assisted in the initial preparations. Particular acknowledgment goes to James H. McGrath for his editorial work on parts of the original manuscript. Quite obviously, the success of this project is most directly attributable to those distinguished philosophers of science who have written the essays contained herein. Finally, James R. Langford, Director of the University of Notre Dame Press, and E. Ann Rice, its Executive Editor, were both enthusiastically responsive in setting up and carrying through a very tight production schedule to have this volume ready in time for Ernan's birthday and for the 1984 Philosophy of Science Association Meeting.

That Ernan McMullin is recognized as an outstanding philosopher of science is evident. While his intellect and wit are well known to all his colleagues in this field, it is his warmth and humanity which most impress those numbered among his large circle of friends. The most eloquent testimony to this has been the enthusiastic response of everyone contacted about contributing to and working on this volume. May Ernan accept this as a small measure of our esteem for his years of leadership and of friendship.

Contributors

NANCY CARTWRIGHT is Professor of Philosophy at Stanford University. Cartwright works primarily in the philosophy of science with special interest in the philosophy of physics. Her recent book, *How the Laws of Physics Lie,* argues for a complex and disparate universe that physics tidies and unifies for simplicity of representation.

JAMES T. CUSHING is Professor of Physics at the University of Notre Dame. For many years his research interests and publications had been in the field of theoretical high-energy physics. He is author of the graduate text *Applied Analytical Mathematics for Physical Scientists* (Wiley, 1975). During the past several years his research and publications have been in the history and philosophy of current physics.

C. F. DELANEY, Professor of Philosophy at the University of Notre Dame, is past Chairman of the Philosophy Department at Notre Dame and current President of the American Catholic Philosophical Association. He is the author of *Mind and Nature* (Notre Dame Press, 1969), and co-author of *The Synoptic Vision* (Notre Dame Press, 1977) and has written numerous professional articles on American philosophy, epistemology, and political philosophy.

ARTHUR FINE is currently Professor of Philosophy at Northwestern University, having previously held that position at Cornell and at the University of Illinois at Chicago. He is a philosopher of science with research interests in the foundations of physics, in topics relating to scientific change, and in the contemporary debates over scientific realism. Formerly a co-editor of *The Philosophical Review,* his work appears in the technical journals in mathematics and physics, as well as in philosophical and historical journals. The present article is part of a long-term research project, using archival materials to help trace the development of Einstein's thought on the quantum theory.

RONALD N. GIERE is Professor of History and Philosophy of Science at Indiana University. He earned his B.A. (Oberlin College, 1960) and M.S. (Cornell, 1963) in physics, and his Ph.D. (Cornell, 1968) in philosophy. During the academic year 1971–72 he was an Associate Research Scientist at the Courant Institute of Mathematical Sciences, New York University. In 1974 he was a visiting Associate Professor of Philosophy at the University of Pittsburgh, and in 1982–83 he was a Senior Research Fellow at the Center for the Philosophy of Science, University of Pittsburgh. Professor Giere is the author of *Understanding Scientific Reasoning* and co-editor of *Foundations of Scientific Method: The Nineteenth Century* and of *PSA 1980*. He has published widely on induction, probability, causality, and statistics, and on relations between history and philosophy of science. He is a member of the American Philosophical Association, the Philosophy of Science Association, the Society for Social Studies of Science, and a Fellow of the American Association for the Advancement of Science.

ADOLF GRÜNBAUM'S writings have dealt with the philosophy of physics, the theory of scientific rationality, and the philosophy of psychology. His books include *Philosophical Problems of Space and Time* (second edition, 1974) and *The Foundations of Psychoanalysis: A Philosophical Critique* (1984). He has served as President of the American Philosophical Association (Eastern Division), and also of the Philosophy of Science Association (two terms). In 1983, a *Festschrift* for him, edited by R. S. Cohen and L. Laudan, appeared under the title *Physics, Philosophy and Psychoanalysis.* A Fellow of the American Academy of Arts and Sciences, and of the American Association for the Advancement of Science, he is delivering the 1984/1985 Gifford Lectures at the University of St. Andrews in Scotland. Currently, he is Andrew Mellon Professor of Philosophy, Research Professor of Psychiatry, and Chairman of the Center for Philosophy of Science at the University of Pittsburgh.

GARY M. GUTTING is Professor of Philosophy at the University of Notre Dame. His primary research areas are philosophy of science, philosophy of religion, and recent Continental philosophy. He is author of *Religious Belief and Religious Skepticism* (Notre Dame Press, 1982), co-author of *The Synoptic Vision: Essays on the Philosophy of Wilfrid Sellars* (Notre Dame Press, 1977), and editor of *Paradigms and Revolutions: Applications and Appraisals of T. S. Kuhn's Philosophy of Science* (Notre Dame Press, 1980).

LARRY LAUDAN is Professor of Philosophy and Science Studies at the Center for the Study of Science in Society at the Virginia Polytechnic Institute and State University. Professor Laudan's major interests in the philosophy of science have centered around the issue of scientific rationality as illuminated by historical case studies of actual scientific practice. His books include *Progress and Its Problems* (University of California Press, 1977), *Science and Hypothesis* (Reidel, 1981) and *Science and Values* (University of California Press, 1984).

EDWARD MACKINNON is Professor in and Chair of the Philosophy Department of California State University, Hayward. He was formerly a Professor at Boston College. In addition to his philosophical training, he has a Ph.D. in physics. His recent publications have been chiefly on the historical development of atomic physics and on the problems of interpreting quantum theory. This work led to his *Scientific Explanation and Atomic Physics* (Chicago, 1982). He is presently engaged in a study of the development of scientific theories. The focus of the research is on the conceptual continuity underlying scientific evolution and revolutions.

HENRY MENDELL is an Assistant Professor of Philosophy at California State University, Los Angeles. The areas of his research interests are Aristotle and the history of early Greek mathematics.

PHILIP L. QUINN is William Herbert Perry Faunce Professor of Philosophy at Brown University. He was educated at Georgetown University, University of Delaware, and University of Pittsburgh; and he has taught at University of Michigan, University of Illinois at Chicago Circle, Ohio State University, and University of Notre Dame. He is the author of *Divine Commands and Moral Requirements* (Oxford: Clarendon Press, 1978) and of numerous articles and reviews in the areas of philosophy of science and philosophy of religion. He currently serves as Secretary-Treasurer of the American Philosophical Association, Eastern Division.

BAS C. VAN FRAASSEN has been Professor of Philosophy at Princeton University since 1982. His current interests are in logic, philosophy of physics and literary theory. He defends antirealism and empiricism in the philosophy of science and a limited use of liberalized probabilistic models in epistemology. His most recent books are *The Scientific Image* (Oxford, 1980) and *Current Issues in Quantum Logic,* edited jointly with E. Beltrametti (Plenum, 1980).

Introduction

GARY GUTTING

When Jim Cushing first suggested that we put together a volume of essays to honor Ernan McMullin on his sixtieth birthday, it was immediately agreed that we did not want a *Festschrift*. That is, we did not want a motley collection of good-hearted efforts to pay one's respects to a friend without throwing away a good paper for a volume no one would read. We wanted instead a set of substantial papers by first-rate philosophers of science. One strategy we briefly considered was to invite papers on a specific topic (e.g., the role of models in science) in the hope of producing a useful specialized anthology. Another was to select a set of topics designed to cover the field of philosophy of science in a comprehensive way, in the hope of producing a valuable survey of the discipline. We rejected both these strategies because it seemed entirely unrealistic to expect a group of outstanding researchers to produce substantial original essays on pre-set topics.

Instead, we decided simply to ask a group of people who are both good friends of Ernan McMullin and important figures in contemporary philosophy of science to contribute major essays on topics that they regarded as central for current research. Our hope—which has been entirely fulfilled—was for a volume that would provide a significant sampling of some of the most important work currently being done in the philosophy of science.

The sample is, of course, not representative of all the major areas in which fruitful work is being done. But it does highlight two areas of central importance. The first is the assessment of the scientific status of various disciplines and theories. Beginning from some general views about the nature of scientific methodology, philosophers of science try to judge whether a given discipline or theory is being (or can be) practiced in a scientific way. Our collection opens with three examples of this enterprise. Ron Giere's "Toward a Unified Theory of Science" provides a remarkably appropriate opening chapter by raising the self-reflective question of the philosophy of science's own scientific status. Adopting a position congruent to Quine's project of "naturalized epistemology", Giere argues that philosophy of science should be practiced as an

1

empirical discipline; that, in other words, it is itself one of those sciences it is dedicated to studying. He defends this construal of philosophy of science from the charge of circularity and illustrates its power by suggesting a unified theory of science expressed in the form of a general theoretical model. This model, he maintains, can be tested on the basis of historical case-studies to which it bears "the standard relationship between theoretical models and empirical data".

Next, Philip Quinn reflects on the status of the "creation science" that has presented itself as a rival to orthodox evolutionary biology. He offers a close critical analysis of the reasoning whereby the judge in a recent case (*Mc-Lean* v. *Arkansas*) concluded that "creation science" is not genuinely scientific. Quinn attacks this argument and argues that the problem with "creation science" is not that it is unscientific but that its scientific claims are insufficiently supported by the evidence.

Finally, Adolf Grünbaum extends his recent work on psychoanalysis with a strong criticism of Paul Ricoeur's view that Freud's theory is hermeneutic rather than empirical. According to Grünbaum, psychoanalysis is an essentially causal account of human actions and, as such, must stand or fall on the basis of the same sort of evidential support that is relevant to natural scientific theories.

All the rest of the essays in our volume deal with questions regarding the ontology of scientific theories. Usually, discussions of such questions focus on the rather narrow issue of the truth of "scientific realism": whether the theoretical entities postulated by advanced sciences such as physics and chemistry actually exist and whether the laws said to govern the behavior of these entities are true or approximately true. The five essays here show how this central but very specific issue connects with a variety of equally important issues that are often ignored in discussions of scientific realism.

Larry Laudan's paper, "Explaining the Success of Science: Beyond Epistemic Realism and Relativism", makes the interesting move of opposing realism not to instrumentalism or fictionalism (its usual contrast terms) but to the sociologically inspired relativism about science that is particularly associated with the Edinburgh School's "strong program" in the sociology of science. This puts the issue of realism into a much broader context than those in which it is usually discussed. Specifically, Laudan invites us to see realism as an alternative to a skeptical rejection of science as a privileged access to reality, with distinctive claims to objectivity and veracity. He argues, however, that, when put in this broader context, realism's defects as an interpretation of science are all the more apparent. Even where we might expect it to show itself decisively superior to relativism—in explaining the success of science—it fails to gain a clear advantage. Realism is in the end no more able to explain the success of science than is relativism; rather, Laudan suggests, the explanation is best provided by an account of science that is anti-relativist but non-realist.

Arthur Fine's "Einstein's Realism" explores the question of what realism means and how it operates not in philosophical reflection on science but in the work of scientists themselves. Focusing on Einstein's espousal of realism (succeeding his early inclination to an operationist or positivist view), Fine provides a careful explication of Einstein's realism and of its relation to issues (such as observer-independence, causality, and the principle of separation) that are central to his critique of quantum theory. His ultimate conclusion is that Einstein's position is in fact much closer to recent versions of anti-realism—especially van Fraassen's constructive empiricism—than to what is usually understood as scientific realism. Nonetheless, Einstein is a "motivational realist"; that is, the ideal of attaining a realistic representation of physical reality is a primary motive for his scientific work.

Another topic relevant to the ontological status of scientific entities seldom raised in contemporary discussions of realism is that of the abstractness of scientific objects. In "What Makes Physics' Objects Abstract?", Nancy Cartwright and Henry Mendell develop an analysis, broadly Aristotelian in character (and influenced by some of Ernan McMullin's recent work), that presents lack of causal structure as the primary factor in an object's abstractness. Unlike Duhem's classic account of abstraction, which is anti-realist in its contrast of the abstractness of the theoretical to the concreteness of the observational, Cartwright and Mendell's account cuts across the theory/observation division and provides the realist with a way to distinguish degrees of concreteness (and, presumably, closeness to reality) of theoretical entities.

Discussions of scientific realism most often focus on the general question Do theoretical entities exist? rather than on questions about the existence of the entities of specific theories. However, the ontological peculiarities of quantum theory raise specific difficulties for the realist and have led some philosophers to argue that, whatever may be the case for other scientific theories, quantum theory simply cannot be interpreted in a realistic way. Since the ultimate structure of matter seems to be best described by quantum theory, this is a major challenge to realism. In his "The Problem of Indistinguishable Particles", Bas van Fraassen discusses a number of problems centering around the question of whether elementary particles governed by quantum mechanical laws violate the principle of the identity of indiscernibles—a principle some philosophers regard as necessarily holding for any real entity. van Fraassen proposes no definitive solutions to the problems he raises and does not explicitly urge the problems as obstacles for realism. Nevertheless, his discussion confronts the realist with the difficult question of (in van Fraassen's words) "how could the world *possibly* be the way physical theory says it is?"

Edward MacKinnon's "Semantics and Quantum Theory" deals with another set of problems posed by quantum theory to the scientific realist. He contrasts the standard approaches to the semantics of quantum theory, which are based on logical reconstructions of the theory, with semantical interpretations (e.g., that of Niels Bohr) based on the theory as it actually functions in scientific work. The former sort of semantics is often thought to yield conclusions inconsistent with realism; but MacKinnon argues that a proper grasp of a theory's ontological significance can be gained only through the latter approach, which, he maintains, supports at least a minimally realistic interpretation of quantum theory.

The essays in this volume, then, present important new ways of approaching some major issues in contemporary philosophy of science. It is also gratifying to note that they deal with issues — particularly, scientific realism — that have been central concerns of Ernan McMullin's work throughout his distinguished career. We hope that he will see this book not only as a warm expression of our admiration of him and his work but also as of some help in his own continuing intellectual journey.

Toward a Unified Theory of Science*

Ronald N. Giere

1. Philosophy of Science Naturalized

Why should anyone be concerned to develop a *theory of science*? The obvious answer is that science continues to be one of the major forces for social change in the modern era. Its importance in modern society thus provides ample motivation, both intellectual and pragmatic, for attempting to understand how it works. And one way, if not the only way, of coming to understand how anything works is to develop a *theory* of how it works.

It is not sufficient, of course, to have strong *motives* for desiring a theory of science. One must have some idea of how to go about achieving this end. My view is that only a *unified* approach to the subject has any chance of success.

Being a complex activity, science has many aspects, including the historical, logical, political, psychological, and sociological. In the actual practice of science, however, these aspects are all bound tightly together. Unlike biology, where the subject matter itself exhibits such relatively clear distinctions as that between cells and organs, science does not exhibit parts or levels reflecting its diverse aspects. The existing divisions in the study of science have been created by the historical fact that the various disciplines which study science developed at different times and in different ways. A successful theory of science will require putting all these aspects together into a unified whole.

Ten or twenty years ago, advocating a unified approach to the study of science would have been a quite radical act. As late as 1960 the divisions between history, philosophy, psychology, and sociology were tightly drawn and strictly enforced. Philosophers of science, in particular, insisted that they studied the "logic of science". Attempts to bring other aspects of science into philosophical discussions were met with charges of "psychologism", "sociologism", or "historicism".[1] Moreover, both psychologists and sociologists pretty much respected the boundaries — confining themselves to the study of "creativity" or the "social structure" of science. Historians pursued mainly intellectual, but sometimes social or psycho-history. Almost no one attempted to make

5

these diverse approaches mesh. Nor is it clear that any such attempt could have been successful. The result was a very disjointed picture of the scientific enterprise.

Inspired by Kuhn (together with Feyerabend, Hanson, Toulmin and others), philosophers, psychologists, historians and sociologists have been breaking down the barriers. With few exceptions (notably Feyerabend), philosophers and historians have proceeded fairly cautiously. Not so the sociologists. Many sociologists, particularly those of the "cognitive" or "interpretive" variety, now refuse to recognize any boundaries at all. Some go so far as to insist that the *only* relevant variables are *sociological* variables.[2] A similar situation is developing in cognitive psychology and artificial intelligence, though psychologists explicitly concerned with science have been less militant in asserting their claims.[3] This leaves philosophy completely out of the picture.

Faced with this competition, philosophers have few alternatives. They can attempt to defend their views on purely *a priori* grounds. They can appeal to a special notion of "conceptual analysis". Or they can try to meet the sociologists (and cognitive psychologists) on the empirical level. For many reasons, some to be presented below, I doubt any defense of *a priori* categories or of special philosophical methods can be successful. The remaining alternative, to adapt a phrase of Quine's, is to "naturalize" the philosophy of science.[4]

Many philosophers of science are somewhat sympathetic with the idea of a naturalized philosophy of science, but few seem completely comfortable with the idea, and some would reject it altogether. It is necessary, therefore, to deal directly both with the discomfort and with the sources of opposition.

2. The Circle Argument

The suggestion that the philosophy of science might itself be treated as a natural science appears clearly in the introduction to *The Structure of Scientific Revolutions*. Kuhn realized that a program to construct "a theory of scientific activity" based on historical data might well be regarded as a "profound confusion". The concepts of confirmation or falsification found in "our usual image of science", he noted, are generally regarded as having to be established *prior to* any empirical investigation of the actual processes of science. They cannot, therefore, be the *results* of such investigation. Nevertheless, he reported, his attempts to apply the usual concepts to "the actual situations in which knowledge is gained, accepted and assimilated have made them seem extraordinarily problematic". We must, he concluded, regard all "images of science" as "parts of a theory" and therefore subject to "the same scrutiny regularly applied to theories in other fields".[5]

Although recognizing the potential logical circle in his procedure, Kuhn wisely declined to face it directly. In this he has not been alone. Philosophers who have embraced the historical turn in the philosophy of science generally pass over the apparent circularity in this approach. But the feeling that the use of historical data to ground conclusions about scientific procedures is inherently circular lies behind much philosophical reluctance to embrace a naturalistic approach to the philosophy of science. To proceed with a naturalistic program, one must break, or otherwise avoid, the circle.

An explicit version of the circle argument would go something like this: To use data to justify any conclusions, one must have some criteria (canons, methods, etc.) for evaluating the relevance of the data. Indeed, even to know what might count as possible data one needs such criteria. Thus, any empirical investigation aimed at discovering the criteria for evaluating evidence would necessarily presuppose at least some of the criteria it was supposedly setting out to discover. So not all the methods of science could be discovered by empirical investigation. Some must be discoverable by *a priori* means.[6]

The discovery of *a priori* justifiable methods for evaluating evidence was a central goal of the major figures in the philosophy of science when Kuhn's book first came out. Carnap, Popper and Reichenbach all sought to discover an *a priori* justifiable method for science.[7] The justification of particular scientific claims would then be secured by reference to these methods. If any such foundationist program were viable, the use of the circle argument against attempts to naturalize the philosophy of science would be vindicated.

But what if such programs are not viable? Quine argued that the foundationist failure to reduce mathematics to logic and semantics to behavior left no alternative but to naturalize epistemology by turning it over to psychology.[8] Other philosophers have been less quick to draw so drastic a conclusion, though almost no one would now publicly defend any foundationist program. If avoiding naturalization required completing a foundationist program, most philosophers would acquiesce. But many have thought that there are other ways out.

One way of attempting to circumvent the circle without pursuing foundationism is to deny that the study of the methods actually used by scientists is empirical in the same way as the methods scientists use in investigating the physical world. Rather, it might be claimed, students of scientific practice are merely seeking to make explicit the methods which they share with natural scientists. The task is *explication,* or *clarification,* not empirical discovery—

and not justification either. For this task, special philosophical methods are required.

Although he would have strenuously denied it, I think that Lakatos provides a primary example of this latter approach. His "methodology of scientific research programmes" is a brilliant synthesis of ideas borrowed from Kuhn, Popper, and Quine.[9] It is quite clear, however, that he did not regard his own theory as a *scientific* theory about how science works. When he came to argue for the superiority of his views over those of Popper and others, he did not apply his own methodology to his own work. Rather, he introduced a special "meta-methodological" rule which applies only to methodological theories. The better methodological theory, he claimed, is that which allows us to "reconstruct more actual great science as rational".[10]

This meta-methodological rule indicates that for Lakatos a methodology is more a theory of *rationality* than a theory of science. Rationality has been for post-positivist philosophers of science what justification was for the positivists. It supplies a *normative* component missing in any purely descriptive, or naturalistic, account. It was just this lack of a normative component that most philosophers of science, including Lakatos, found most objectionable in Kuhn. They wanted a clear distinction between knowledge and mere belief—a distinction that has traditionally been made in terms of justification.

Framing the issue in terms of rationality rather than justification, however, does not eliminate many of the issues faced by the positivists. In particular, one can still ask whether a proposed account of rationality really is such. Lakatos's only answer can be that great science is by definition rational. So any correct account of great science is automatically an account of scientific rationality. This is why I think that Lakatos's program really is to be viewed as one of explication.

The question arises whether explication based on anything like Lakatos's meta-methodological rule could be sufficient. I think not. Even though it was not intended as a scientific theory, Lakatos's methodology suffers from the severe underdetermination faced by any theory based on a fixed, and finite, set of data—the cases of "great science". I do not mean underdetermination in the trivial sense that any theory is *logically* underdetermined by finite data. Astronomy provides the relevant sort of comparison. So long as one has only to account for the apparent motions of the planets as viewed from the Earth, it is difficult to choose between geocentric and heliocentric astronomy. If we add a specific dynamics, however, the choice may be determined. Newtonian dynamics implies, for example, that a vastly more massive body must be the dynamic center of a gravitational system. By requiring only that a methodology permit a favorable reconstruction of existing great science, Lakatos's methodology is in the position of a purely kinematical astronomy. It is very difficult to distinguish it from other accounts of the same data — as Laudan's rival methodology demonstrates.[11]

I do not see how to avoid this objection without using the actual cases of good science not as a criterion of rationality, but as part of an *empirical test* of one's account of science. But this move would again face the circle argument. Rather than trying to slip out of the circle, perhaps it is necessary to find some way to cut it without resorting to epistemological foundationism.

3. An Evolutionary Perspective

The circle argument is a version of classic arguments concerning the justification of induction. Rather than requiring a prior justification for a particular factual statement, it demands a prior rule, or method. This relationship with the problem of induction may partly explain the power of the circle argument within the philosophical community. It may also discourage one from attacking it head on. The task, however, is not to solve the problem of induction, but to explain why it is not, after all, a serious problem.

Since the seventeenth century, epistemological investigations have begun with a solitary individual facing his own immediate, conscious experience. One remembers Descartes sitting by his fire and David Hume in his study. The problem for both rationalists and empiricists was how to proceed from there. This was a reasonable response to the intellectual climate of the seventeenth century. It was important for these thinkers to show that the epistemological authority of science was at least comparable to the authority of the church.

The same response is not reasonable in the twentieth century. Thanks to Darwin and a century of further research in evolutionary theory and genetics, we now know that no humans ever faced the world with only their subjectively accessible experience to guide them. Our perceptual and other cognitive capacities, which are poorly reflected in our subjective experiences, are very well adapted to the environment in which we evolved. There is no longer any need to wonder why we are not systematically deceived by our environment, or why we associate colors and motions with material objects, or why we perceive things as being in a local Euclidian space-time, or why we possess an empathetic understanding of our fellow humans. If we had not evolved these capacities, we would not be here.

The philosophical skeptic, of course, replies that to invoke evolutionary theory simply begs the question. Evolutionary theory is a fairly advanced form of scientific knowledge, and the problem is to justify that knowledge from something one can experience directly. But here it is the skeptic who begs the question. Three hundred years of modern science and over a hundred years of biological investigation have led us to the firm conclusion that no humans have ever faced the world guided only by their own immediate experience (perhaps supplemented with memory of previous experiences). The skeptic asks us to set all this aside in favor of a project that denies our conclusion. What reason could there be to embark on a project which we now know has little relevance to the actual processes by which humans learn about their environment? Only *scientific* evidence that we are in fact grossly mistaken in our understanding of evolutionary biology, neurophysiology, and so on. The skeptic offers merely the *logical* possibility that we are wrong. But since it is impossible to close this logical gap, there is no reason to try.

There is, then, no compelling reason to embark on either an empiricist or a rationalist project of justifying induction. Nor need we be dismayed that several centuries of thought have not produced a solution. There is no reason to think that any solution is even possible, let alone necessary.

From our post-Darwinian perspective we see that humans must be fairly well adapted to their environment, but only insofar as their adaptation promotes the production of viable offspring. That is how evolution works. The evolutionary perspective does not by itself tell us how we humans have been able to use our evolved capacities to acquire our vast knowledge of atoms, stars, and nebulae, of entropy and genes.[12] That is the job of an empirical theory of science.

There remains the more general philosophical objection that we must be able to distinguish between conclusions that are somehow "justified", or "rational", and those that are not. There is more to epistemology, it is held, than merely the psychology (and even the sociology) of belief. This general misgiving about naturalistic epistemology or philosophy of science cannot be met by theoretical argument alone. One must provide a naturalistic account of the distinction. I will outline such an account below.

To conclude this section, it should be noted that my appeal to evolutionary theory is far more limited than that of numerous advocates of "evolutionary epistemology". These authors have usually taken evolutionary theory itself as a model for the development of scientific knowledge. I doubt it is a very good model, and I shall suggest a far different account. We are all agreed, however, that the issue is an empirical one, to be settled by scientific procedures, not by philosophical argument.[13]

4. Constructive Realism

There is now widespread agreement among students of the scientific enterprise that our accounts of science must include the *cognitive content* of the sciences. In particular, *theories* must play a prominent role in any theory of science. Since philosophers have been preoccupied with the nature of theories, one might hope that this work would be helpful in constructing a unified theory of science. Such is unfortunately not the case.

Taking the foundations of mathematics as their model, the Logical Empiricists portrayed scientific theories as axiomatic systems ideally formulated in a formal language such as first order logic or set theory. Attempts to reconstruct particular bodies of theory utilizing the prescribed logical resources have mostly provided some interesting puzzles for logicians. They have not measurably improved our understanding of the theories in question. More general philosophical questions about theories, such as "the problem of theoretical terms", mainly reflect the goals of classical empiricism. As Kuhn and others recognized, these problems are not mirrored in the actual practice of science.

If the Logical Empiricist account of theories was too narrow, the alternative offered by post-positivist philosophers and social scientists is too broad. This was particularly true of Kuhn's original "global paradigms", which included everything from metaphysical ideas to instrumentation. His later distinction between an "exemplar" and a "disciplinary matrix" was a step in the right direction. I think an even more fine-grained account is necessary to do justice to actual science.[14]

Drawing on some recent work in the philosophy of science, the basic ingredient in my account of theories is the notion of a "theoretical model".[15] I intend this notion to capture much of what scientists themselves intend when they talk about "models". Of course it is neither necessary nor sufficient that concepts in the theory of science should overlap with concepts used by scientists. But overlap is desirable where possible because it facilitates communication with one's subjects who are, after all, highly intelligent people often possessing fairly well-developed theories of their own about what they are doing.

A theoretical model (hereafter 'model' unless otherwise noted) is an idealized, abstract object or system of objects. Moreover, models are *socially constructed* objects. They are constructed using the linguistic resources of the community. Typically these linguistic resources include everyday language plus specialized concepts, including those of mathematics. It would be a mistake, however, to *identify* models with language. We want to regard several scientists as using the *same model* even though they speak *different languages,* say

English and Japanese. More important, we want to allow that individual scientists may possess a greater understanding of a model than they could in fact put into words. That is, we want to allow "tacit knowledge" even at the most theoretical levels. Moreover, we want to allow the possibility that a group of scientists might possess a model that no one member could fully articulate.

Well-known examples of models in the twentieth century include the Bohr model of the atom, particulate models of inheritance, plate tectonic models of the earth, and functional models of society. As these examples illustrate, models are typically models *of* something or other. I will return shortly to the relationship between a model and what it is a model of. Here I want only to emphasize that a model is not to be *identified* with that of which it is a model.

One feature that makes models very useful in science, and thus makes 'model' a useful concept in the science of science, is that models come in various degrees of *specificity*. At their most non-specific, theoretical models include things often labeled "metaphysical". Atomic and continuous models of matter are rival, though highly non-specific, types of models. A slightly more specific pair would be contact models of mechanics and action-at-a-distance, or field, models. Descartes's vortex models and Newton's gravitational models are more specific versions of these general types. A Newtonian model of a two-particle system with non-zero angular momentum is a still more specific model. Finally, if we add particular values for the masses, initial velocities, and positions, we obtain a *maximally specific* version of a Newtonian model of a two-body system. Note that being maximally specific is a relative concept. There is no such thing as a maximally specific model, period, It is only a model *of a designated type* that can be maximally specific.

For models of a given type, the less specific do not determine the more specific. There are many ways of filling in a highly non-specific model to achieve a highly specific version of that model. The relationship between non-specific and specific models, therefore, should not be confused with the relationship between *general* and *particular,* as in the relationship between general laws and particular instances. That is a very different and, from my point of view, a much less interesting type of relationship.

Let us return to the relationship between a model and that of which it is a model. This requires a new concept, that of a *theoretical hypothesis.* Here again there is intended to be considerable overlap with the use of this concept by scientists themselves, though my usage is more systematic. A theoretical hypothesis, on my account, is a statement asserting that some designated real system (or class of systems) *is similar to* a given model in specified *respects* and to specified *degrees*. The following, for example, is a theoretical hypothesis: The positions and velocities of the earth and moon in the earth-moon system are very close to those of a two-particle Newtonian model (with specified initial conditions). A less stilted formulation, one closer to how scientists actually talk, would be: The earth and moon are, to a high degree of approximation, a two-particle Newtonian system. This latter formulation tends to blur the distinction between the theoretical model and the real system, a distinction which a theory of science should, I think, keep quite sharp. It also fails to distinguish respects and degrees, lumping both into the vaguer notion of "approximation". But so long as the distinctions are understood, the more relaxed formulation is acceptable.

Like models, hypotheses come in various degrees of specificity. Indeed, any theoretical hypothesis reflects the specificity of the model to which it refers. But hypotheses, unlike models, can also be more or less *general* in the sense of including more or fewer real systems of various kinds. It is primarily quite general hypotheses, or hypotheses applying to particularly important objects, that tend to go by the name of 'theories'. Consider, for example, Newton's theory of celestial mechanics, Mendel's theory of inheritance, or the plate tectonic theory of the earth. On this account, then, the difference between a 'hypothesis' and a 'theory' may be largely honorific.

In terms of traditional philosophical categories, a hypothesis (or theory) can be true or false. It makes no sense to call a model true or false since models are not sentences or statements. My account of theories, therefore, can legitimately be labeled 'realistic'. But it is a modest, *constructive realism* which acknowledges that models are human constructs that could not be expected ever fully to capture the richness and detail of any real system. Moreover, it is a realism that lays no stress on *universality*. Of course we seek models with wide applicability, but universal applicability is not a necessary condition for scientific status. And sometimes, as in human psychology or geology, it is the human significance of the subject matter, not the range of applicability, that makes the theory worthy of pursuit.

The fact that hypotheses may be more or less specific explains the wide divergence of opinion regarding the *falsifiability* of theories by empirical data. Logical Empiricists, including Popper, made falsifiability a necessary condition for a hypothesis to be scientific. Kuhn and later philosophers, with considerable historical data, argued that theories are never falsifiable. The fact is that highly *specific* hypotheses can quite easily be refuted, but that highly *non-specific* hypotheses can be refuted only with great difficulty if at all.

Consider any ordinary pendulum, the kind found on some clocks. And consider the textbook Newtonian model which exhibits a pendulum as a type of simple harmonic oscillator. Take the respect of interest to be how long the pendulum continues in motion once set freely swinging. Since a simple harmonic oscillator is a conservative system, it continues oscillating forever. Our ordinary pendulum sooner or later comes to rest. The hypothesis that this pendulum is similar to the model in the respect indicated is refuted. Here Popper is right.

But now consider the hypothesis that the motion of our real pendulum is captured by *some* Newtonian model. This is a far less specific hypothesis. Failure of the real pendulum to fit the simpler model does not imply that there is not some more complex Newtonian model that does the trick. Scientists might give up trying, but in the absence of some rival theory implying the impossibility of any such model, it would be very difficult to prove experimentally that one could not be found. This is the kind of situation emphasized by post-positivist scholars.

A theory of theories is only a part of a theory of science. Nevertheless, the fact that the above account makes sense of a variety of insights about scientific theories provides hope that it is a step in the right direction.

5. The Similarity Argument

At first sight it seems that even a modest constructive realism is incompatible with relativism, and with the interest models of science now popular among British sociologists of science. Yet there is a widespread argument that would place constructive realism snugly in the relativist household. I wish to refute this argument and thereby to escape relativism.

The argument in question is a purely theoretical (not to say philosophical) argument which is usually attributed to the philosopher of science, Mary Hesse, though it can be found as well in Kuhn's early writings. It does not appeal to sociological data. And although it is usually stated in terms of perceptual categories, it should apply with even greater force to theoretical categories.[16]

On my account, the relationship between a model and the world is one of *similarity* in various respects and degrees. Now, as a matter of logic, anything is similar to anything else in some respects and to some degrees. Any theoretical hypothesis, therefore, is vacuously true. To make it non-vacuous, one must specify respects and degrees. But then its truth or falsity depends entirely upon this specification. And agreement on its truth or falsity depends on prior agreement on the respects and degrees. But these are not given either by the character of our theoretical models or by the nature of any real system. If there is agreement, therefore, it can only be because there was a prior agreement on respects and degrees which is *socially* sanctioned and/or enforced. Thus, which hypotheses are called true and which false depends entirely on social agreements which are totally independent of our models and of how the world really is.

Now I agree with the first part of this argument. The truth value of a theoretical hypothesis is indeterminate without some specification of respects and degrees. Nor would I deny that agreement on respects and degrees is to some extent socially determined. The question is how this determination takes place and thus how much freedom there is for the operation of social or other interests. There is, I think, far less freedom than supporters of the similarity argument have supposed.

As is suggested by its more usual formulation in terms of perceptual judgments, the similarity argument is rooted in pre-Darwinian empiricist epistemology. It assumes the Humean view that there is no natural connection between any two "impressions". The only connections are those we impose. Yet the effect of evolution on our sensory apparatus has been particularly strong. Animals are capable of incredibly fine discriminations among objects in their environment without benefit of social conventions. And so—being no more than fairly intelligent talking primates—are we. There is, for example, no human group among the hundreds that have been studied, which does not distinguish red from green, though many do not distinguish pink from red. For at least some perceptual judgments, therefore, the fact of widespread agreement does not require a social explanation. The explanations of evolutionary biology are sufficient.[17]

The direct force of the evolutionary argument for theoretical judgments is obviously far less strong. But it is not empty. Animals actively explore their environments and construct fairly complex representations of their surroundings. Our ability to construct theoretical models to represent our environment is continuous with the abilities of lower animals—indeed, it makes use of some very similar neural mechanisms. Thus, although our representations gain much from our linguistic abilities, the goal of representing our environment, which we share with our animal ancestors, may still provide some natural constraints on our judgments even of theoretical similarities.

Regardless of the direct influence of our biological capabilities on our theoretical judgments, these judgments are also constrained by their connections with perceptual judgments. And perceptual judgments of similarity have some biological basis. I will now try to illustrate how theoretical judgments are, at least sometimes, actually made. My account explicitly maintains the connection between human and animal behavior. It may be regarded as a naturalistic counterpart of philosophical theories of justification or rationality.

6. The Revolution in Geology

When *The Structure of Scientific Revolutions* was first published, the earth sciences were in the midst of a scientific revolution. It took historians and philosophers a decade or more to begin studying this revolution—sociologists and others have yet to do so. It is a good case for my purposes.[18]

As late as 1950, the standard theory of the earth was based on the model of a molten sphere, suspended in space, which, upon radiating heat into space, cools and contracts. Fairly detailed models of this type existed already in the nineteenth century. The main features of the earth's structure—continents, oceans, mountains, and so on—were explained in terms of the contractionist model. In particular, it is a consequence of this model that the continents have always been in more or less their present locations. The model contains no lateral forces to move continents horizontally over the surface. Moreover, the material making up the continents is generally softer than that of the ocean floors, having floated to the top during the original cooling. A softer material cannot move through a harder one, even if there are forces pushing it.

Not everyone, of course, was completely happy with the contractionist model. There are many features of the earth that do not fit well in any contractionist scheme. The coastlines of North and South America, for example, exhibit an apparently remarkable match with the coastlines of Europe and Africa. Also, there are some very similar plants and animals on different continents, such as South America and Africa. Both of these facts, and many others, could easily be explained by a model in which the continents move *horizontally* over the surface of the earth. Indeed, in 1915 the German geologist, Alfred Wegener, published a detailed model which allowed the possibility that North and South America were once attached to Europe and Africa. Their subsequent breaking off and drifting apart explains the match in coastlines and the similarities in flora and fauna. Wegener's theory of the earth attracted a fair amount of attention over the years, most of it negative. In 1950, anyone seriously suggesting that a drift model might be correct ran the risk of being labeled a crackpot.

Two things happened in the 1950s which made the revolution of the

1960s possible. Neither initially had much to do with global models of the earth's structure. One was the study of paleomagnetism, that is, the study of the history of the earth's magnetic field. By the early 1960s there was accumulating evidence that the earth had reversed its magnetic poles a number of times over the past four or five million years. Most of this evidence was in the form of core samples at various places on the surface of the earth, such as volcanic areas, which would preserve the magnetic orientation of originally molten materials. The reversals seemed to be quite rapid, taking only a few tens of thousands of years, and lasting around a million years. But there also appeared to be several short reversals lasting only a hundred thousand years or so. The pattern of reversals thus has a fairly distinct "signature".

The second development was the discovery of large ridges on the ocean floors, one roughly down the middle of the Atlantic Ocean and others in the eastern Pacific. These ridges have a trough running down the center and are roughly symmetrical to either side. Moreover, the troughs were found to be generally warmer than the surrounding sea floor. In 1960, Harry Hess, a respected geologist, speculated that the ridges might be due to currents of molten material which rise up below the ridges and spread out *horizontally* forming the ocean floor on either side. Hess was quite aware that his hypothesis implied sea-floor spreading which, in turn, provided a mechanism for continental drift. The continents need not push through the ocean floor; they could ride along on top as the floor itself moves.

Among people who heard Hess present these ideas was Fred Vine, a graduate student in petrology at Cambridge. During the next few years, Vine, together with his supervisor, Drummond Matthews, himself a recent Cambridge Ph.D., put together Hess's idea of sea-floor spreading with the results on geo-magnetic pole reversals. They reasoned that the originally molten material forming the new part of the ocean floor on either side of a ridge should also preserve the magnetic orientation of metallic particles. Thus one should find the history of the pole reversals recorded as oppositely magnetized strips of ocean floor parallel to the ridge. And these strips should produce the same "signature" as found in terrestrial samples. This hypothesis was conceived independently by Canadian geologist Lawrence Morley. It is now commonly referred to as the Vine-Matthews-Morley (VMM) hypothesis. Vine and Matthews published their hypothesis in *Nature* in 1963.

During the next two years, Vine and others analyzed existing data on magnetism of the sea floor near ridges. The data seemed to bear out the VMM hypothesis, but it was generally too "noisy" to be really convincing. And not many people were convinced, including scientists at the Lamont Geological

Observatory, one of the three leading geophysical institutions in the United States. Lamont's founder and director, Maurice Ewing, was one of the best-known opponents of drift models. His opinion, not by accident, was shared by most of his associates. But Lamont's research capabilities, including several research vessels, were among the best in the world, and since 1945 they had gathered a vast store of data, including core samples from the ocean floor.

In the fall of 1965, one of their research vessels, the *Eltanin*, was making a survey, including magnetic readings, across the Pacific-Antarctic Ridge in the southeastern Pacific. About the same time, stimulated by the work of Vine and others, some researchers at Lamont began studying their sea-floor core samples for magnetic reversals. The data from both projects became available early in 1966 and showed a remarkably *clean* match between the two "signatures". The pattern of magnetic stripes, showing a clear mirror-image symmetry across the ridge, was itself particularly striking. Lamont's data was presented at the April meeting of the American Geophysical Union. Almost everyone was immediately convinced that sea-floor spreading was a reality. This included almost everyone at Lamont except Ewing, who held out for several years.

The above is but a sketch of a complex history, and my selection of events is clearly biased by my own theory of science. But I offer it not strictly as evidence for my view, but mainly as an illustration. The public record, in both primary and secondary sources, is now sufficiently great that anyone wishing to dispute my interpretation need only consult that record.[19]

7. Naturalistic Theory Choice

How do we account for the rapid switch by earth scientists from contractionist to drift models of the earth? Logical Empiricists would say that there was a "logical relation" between the hypotheses and the new evidence that made it *logical,* and therefore *rational,* to accept the drift hypothesis. An ideal Carnapian inductive logic, for example, would assign a high logical probability to the drift hypothesis and a very low logical probability to the contractionist hypothesis. The original Popper would say that the contractionist hypothesis was refuted by an application of *modus tollens* while the drift hypothesis survived a severe test.

The Kuhn of 1962 would have said that the older, static paradigm underwent a crisis due to the recognition of many anomalies. What emerged from the revolution was a new paradigm based on the hypothesis of mobile continents. It was not a matter of logic or rationality. It was simply a change in paradigms.

Lakatos would have said that the drift research programme suddenly became much more "progressive" than the static program, which began rapidly to degenerate. Scientists were *rational* to switch to the more progressive programme. Laudan would say that the "problem solving effectiveness" of the drift tradition became suddenly much greater than that of the static tradition. It was *rational* to switch to the tradition with greater problem solving effectiveness.

Recent cognitivist sociologists would tell a very different story. Interest theorists would say that it suddenly became in the professional and/or social interests of scientists to switch views. Others would appeal to a process of *negotiation* by which the "reality" of continental drift was constructed.

It is an illuminating exercise to evaluate these theories of science in light of this case, but I cannot do that here.

I agree with Kuhn and recent cognitive sociologists that there is no special scientific rationality which could justify any such scientific behavior. Not only is there no logic of discovery, there is no logic of *justification* either. It is not a matter of determining which hypothesis is rationally supported by the evidence and then simply doing what is rational. Rather, as many sociologists have urged, we must look at the actions of scientists as we would actions by humans in any other social context.

I propose that we focus on the activity of *choice*. Scientists make many choices in their careers and even in their daily scientific lives. Sometimes they must choose between competing theories, which is the case of interest here. So far most sociologists and some philosophers could agree. Our disagreement is over the particular model of choice we should employ.

Although apparently not prominent in the sociological tradition, there are well-developed models of decision making that can be applied to the scientific context. Perhaps decision-theoretic models have seemed too abstract, and too normative, for sociological taste. Perhaps, also, sociologists have been more concerned with the *process* of decision making than with the more abstract *structure* of the decisions made. Yet attention to the structure of decisions can be very revealing. I admit that decision-theoretical models are quite idealized, but no more so than models in other sciences. The real issue here, as in other sciences, is whether these models capture enough of the structure of the situation to be informative. That cannot be decided *a priori*.

Let us consider, therefore, the situation of professionally active geologists who were present at that famous 1966 meeting of the American Geophysical Union, but who were not personally involved in the research at issue. These subjects are not disinterested, however, because the overall tectonic structure of the Earth is relevant to their own fields and to their own research. Moreover, we must assume that our subjects start off with a bias toward contractionist models. Those were presumably the models they were taught as students and the models that underlay their current research. Yet in light of discussions at that meeting, these geologists could not fail to consider the question, perhaps explicitly for the first time, whether drift models might not, after all, provide a better representation of the actual tectonic structure of the earth.

The simplest model of this decision has the structure of a two-by-two matrix. There are two possible choices: (a) Continue taking contractionist models as basically correct. (b) Take drift models to be more nearly correct. Correspondingly, the actual tectonic structure of the earth is (i) roughly contractionist or (ii) roughly drift-like. This creates four possible outcomes of the scientist's decision. In two possible outcomes the decision is correct and in two it is mistaken.

Given the assumed professional bias, our scientists most prefer the outcome in which contractionist models are correct and they (rightly) continue taking them to be so. Less preferable, but still satisfactory, is the outcome in which drift models are correct and the decision is to decide in favor of drift models. The worst outcome would be to decide in favor of drift models when the contractionist tradition is in fact correct. It would be better, but still not very good, to stick with contractionist models when drift models are in fact correct. The resulting matrix is exhibited in figure 1.

The preferences represented in figure 1 are the decision theorists' representation of the scientists' interests or values. The indicated ranking expresses only the roughest measure of these values. It turns out that a rough measure is sufficient.

I have not actually interviewed or otherwise empirically attempted to

Figure 1. Decision Problem for Geologists in 1966.

	Drift Models Approximately Correct	Static Models Approximately Correct
Adopt Drift Models	Satisfactory	Terrible
Retain Static Models	Bad	Excellent

ascertain that this is indeed the preference ordering that would have been picked by a majority of scientists meeting my description. I am simply using my own scientific intuitions regarding the situation as I understand it. What I say may thus be challenged on empirical grounds. I expect, however, that most students of science would agree at least with my rank ordering of the outcomes. In any case, I doubt anyone currently has data that would refute my ranking. So, at least for the moment, we are stuck depending on our own scientific intuitions about these preferences.

It may be objected that scientists do not, or even cannot, place any particular value on making the objectively correct decision—that is, choosing the model that is in fact most similar to the real structure of the earth. What matters is choosing the model that follows the consensus of the profession. For many scientists much of the time, this is correct. But we are considering a case in which the consensus is in the process of reformation. In such cases the emerging consensus will be a function of the number and influence of scientists who choose a particular model. In these cases, I claim, individuals are trying to choose the objectively correct model.

An alternative would be to claim that they are all trying to choose the model that *will* emerge as the consensus model. Each is trying to guess what the others will do. This is possible, but does not fit at all with the way scientists in fact talk about what they are doing. Nor does it square well with their actions, at least on the surface. To substantiate this alternative one would have to explain how and why scientists are doing something quite different from what they say they are doing, and give independent evidence that this alternative explanation is correct.

What, then, does decision theory tell us about the matrix in figure 1? Not very much. The only applicable decision rule is minimax, which enjoins us to choose the option with the highest guaranteed minimum payoff. In this case, that means sticking with contractionist models, since this choice guarantees at least the next-to-lowest-ranked outcome. Choosing drift models could yield the worst outcome. Since most of the scientists in question chose the drift model, they obviously were not following a minimax strategy—assuming the matrix in figure 1 correctly represents the decision as they saw it. This is not surprising since minimax has never provided a *descriptively* adequate account of very many cases of human decision making.

It is clear in any case that figure 1 does not tell the whole story. The new evidence of sea-floor spreading from the *Eltanin* and from core samples played a major role in the decision, and this evidence is not represented in the matrix of figure 1. What role, then, did the evidence play?

As in statistical decision theory, the evidence operates indirectly on the choice of a model by means of a *decision rule*. The decision rule in this case was the obvious one: If the VMM hypothesis is verified, choose drift models as being more nearly correct; if the VMM hypothesis is refuted, maintain adherence to contractionist models. I do not claim that anyone explicitly formulated this rule, though someone might have. I do claim that the relevant scientists were acting in accord with this rule. Why, then, is it the obviously correct rule?

The evidence verifying the VMM hypothesis was just that expected if a drift model were correct, and it would not have been expected according to any contractionist model ever proposed. Invoking probabilistic notions, the evidence was highly probable assuming a drift model and highly improbable assuming a contractionist model. This is not to say that anyone could calculate such probabilities. Rather, the judgment that the one probability was high and the other low was shared by just about everyone professionally familiar with the issues. Nor is this to say that there was no *logically* possible contractionist model that would make the evidence highly probable. It is just that no one could think of any modification, consistent with known facts, that would do the trick.

Looking at the matrix in figure 1, we see that following the indicated rule makes it highly probable that one will end up with one of the two satisfactory outcomes and highly improbable that one will end up with either of the unsatisfactory outcomes. The rule, therefore, is an example of what Herbert Simon calls a "satisficing" strategy.[20] As an empirical matter, humans and many animals tend to be "satisficers". That is, in situations of uncertainty they look for strategies that have a high probability of yielding a satisfactory payoff no matter which of the possible contingencies is realized. If such a strategy can be found, they generally take it. This is not a principle of rationality, but merely a description of goal-seeking activity in the face of uncertainty. The operation of such a strategy is plausible even in animal experiments.

Assuming, then, that the geologists at the 1966 AGU meeting were satisficers, they would adopt the suggested rule and their choice of the drift model followed as a matter of course.

It should be noted, finally, that the above analysis is highly robust both with respect to the relative values assigned to the various outcomes and with respect to the probabilities of the evidence on the alternative models. Scientists could have fairly divergent opinions on these matters and still make the same choice.

8. A Function for Experiments

Doing experiments has been one of the most obvious features of science since the seventeenth century. Yet few current theories of science explicitly address the question: What is the function of experiments in science? Popper, with his emphasis on severe testing, is one of the few empiricist philosophers to emphasize experiments. Others apparently regard experiments merely as a source of evidence to be evaluated by a proper logic of science. Kuhn has dealt seriously with the issue, but few later theorists, whether historian, philosopher or sociologist, have gone further.[21] The only recent novel suggestion is that experiments are a powerful "bargaining chip" in negotiations concerning scientific reality.[22]

My analysis provides a simple and immediate answer to the question. A function of experiments is to aid in the process of deciding among competing theoretical models. This is, of course, not the only function experiments may serve, but it is a major function. When there are only two alternative models, the best experiments, as far as fulfilling this function is concerned, are those which provide a satisfactory decision strategy for most scientists (or the most influential scientists) for whom the choice is professionally relevant. In this special case, a satisfactory experiment is one that promises to generate data which is very probable if one model is correct and very improbable if the other is correct. Such an experiment will fail to be decisive only for scientists who value other things much more than they value being right.

My account may be viewed as a naturalistic, decision-theoretic version of the old doctrine of "crucial experiments". But there is no suggestion that such experiments are *logically* conclusive. To be crucial in my sense requires many contingent preconditions. First, the alternative models must be explicitly characterized. Logically possible models which no one has described are irrelevant to the decision. Second, there must be general agreement that a specified experimental result would be highly probable on one model while being highly improbable on the other model. This requires widespread agreement on a considerable body of background knowledge. Such agreement does not always exist. But sometimes it does. Sometimes there may even be agreement on the potential decisiveness of a test *before* the results are in. Finally, the data must be sufficiently clear that most participants will agree that the expected result really did occur.

These conditions were all present in the case of plate tectonics, and their presence contributed substantially to the speed with which drift models were accepted. Moreover, there are many factors in this case which make interest or negotiation models difficult to apply. The interests of the community were strongly biased in favor of contractionist models right up to 1966. The reception of the VMM hypothesis in 1963 was decidedly negative. The champions of drift models prior to 1966 were not powerful figures in the field—Vine was still a graduate student. Finally, the decisive data were obtained by a research group publicly opposed to drift models. When this data came in, all the sociological factors were simply swamped. The appropriate decision was obvious to almost everyone.

9. The Laws of a Theory of Science

At the end of their already classic study of the development of radio astronomy in Britain, Edge and Mulkay list fifteen "factors in scientific innovation and specialty development" which are exhibited in at least one of a half dozen studies. Several of these factors are exhibited in all six of the studies surveyed. The factors cited include such things as "marginal innovation", "mobility", and "creation of new journal".[23]

Although they do not describe their remarks this way, Edge and Mulkay may be viewed as looking for some "laws" of innovation and specialty development. This interpretation suggests the broader question: What might the laws of a theory of science be like?

It is helpful to consider the role of evolutionary theory in biology. In the nineteenth and early twentieth century, biologists looked for the "laws of evolution" among existing species and in the fossil record. Bergmann's Law, for example, states that for species of warm-blooded vertebrates, races living in cooler climates are larger than races living in warmer climates. Contrary to what some empiricist philosophers have claimed, the trouble with such laws is not that they fail to apply throughout the universe, but that they hover between falsity and vacuity even when restricted to life on earth. If stated with sufficient precision to be informative, there are always some exceptions, e.g., burrowing mammals. If stated with sufficient vagueness to avoid exceptions, they cease to be usefully informative. The reason these "laws" have such a dubious status is that they are attempts to generalize over what are partly *initial conditions* (including the environment) of evolving systems of organisms. At best these statements express rough statistical generalizations reflecting the average selective forces of different environments on similar populations of organisms.

A better place to find the laws of evolution are in the descriptions of the "theoretical models" used in formulating the hypotheses of evolutionary theory. The most likely candidates for such laws are, very briefly, (i) random variation of heritable phenotypic traits, and (ii) differential fitness, relative to the environment, of organisms with different phenotypic traits. Any population of organisms satisfying these laws will evolve, provided only that the environment is not so hostile as to preclude survival. The theory of evolution is the general hypothesis that all (or at least most) species of organisms on Earth evolved by processes exhibiting a structure characterized by these laws. On this characterization, evolutionary theory does not tell us much about how specific species will evolve in particular environments. But neither does it preclude the discovery of rough generalizations covering limited evolutionary situations.

The "laws" of specialty development suggested by Edge and Mulkay's factors are like the nineteenth-century laws of evolutionary development. They provide at most some rough generalizations covering a very restricted class of cases. The problem, as in the biological case, is that the phenomenon of science is too variable at the level of the types of factors surveyed. Just as nature may solve an evolutionary problem in many different ways, so specialty development may occur in many different ways. A theory of science, like a theory of evolution, must be based on models which capture the "deeper structure" of the subject matter.

As a rough attempt to characterize a general model of scientific development, I offer the following: A community which can support (i) the creation of theoretical models specific enough to be tested empirically, and (ii) the design and execution of experiments which will permit a satisfactory decision as to whether the proposed models are in fact sufficiently similar to the real systems of interest.

Here one must be careful not to overinterpret the biological analogy. Models are not produced by random variation. Creating models is a highly purposeful activity, and the kinds of models scientists are likely to create will depend strongly on their prior knowledge of the systems being modeled. Moreover, although testing may be regarded as a process of selection, there is nothing resembling differential reproductive potential which constitutes the biological concept of fitness. In addition to being biological, our model of science must ultimately be both *cognitive* and *social*.

Although the above model of science is highly non-specific, anyone familiar with the history of science will be able to suggest examples in which one or another of my two "laws" appear to be violated. I cannot argue such cases here. Rather, at the risk of evoking still more counterexamples, I will suggest a further, contingent factor that seems essential for the development of any scientific specialty or theoretical tradition. This is that the community experience some degree of *success*. That is, some of its models test out positively. This is not to say that all attempts must be successful. Far from it. But without some clear successes it is difficult to see how a specialty can maintain itself and recruit new members. But whether this suggestion will withstand comparison with the cases remains to be seen.[24]

10. The Place of History

Much of the extensive literature on the relationship between history and philosophy of science has been inspired by the need to reconcile a *normative* discipline (philosophy) with a *descriptive* discipline (history). A naturalized philosophy of science should have much less trouble sorting out its relationships with history.

Let us begin by considering the relationships between other sciences and the histories of their respective subject matters. Though physics is widely regarded by humanists and social scientists as somehow atemporal, this view rests on a gross misunderstanding. It is true that physical models represent *all possible* histories of a system. They thus describe the causal structure of the system apart from its actual history. Once initial conditions are provided, however, the result may be a single history, hopefully one corresponding to the actual history of a particular system.

Newtonian models of the solar system, for example, may be used to determine the history of the positions of the planets using current positions as initial conditions. In this way the relative positions of the planets at the birth of Christ or the year 2000 may be determined. That is the beauty of deterministic models. But even in physics the relationship between theory and history is not all one way. Historical records, say of the position of the moon, do not square with simple models of the earth/moon system. The records show that more complex models, ones including the effect of tides, for example, are required.

Physical models can do a very good job of representing relatively stable systems with discernable initial conditions. They are often very poor at representing origins or other major changes in structure. That is why there is no good Newtonian model of the origin of the solar system. Even if the process has been totally classical, which it probably was not, the initial conditions are now largely beyond our reach. This means that some facts about the actual makeup of the solar system will remain irredeemably historical. That's just the way it was earlier, and we now have no way of discovering why.

In biology the relationship between our models and the actual history of biological systems is much looser, if only because evolutionary models are essentially *stochastic*. A population may evolve in a particular direction simply because one mutation occurred by chance before another. Biological theory provides no further explanation of the resulting development. The explanation of some aspects of a population's development, therefore, remain irremediably historical.

In biology there is strong interaction between evolutionary theory and the historical record, as studied, for example, by paleontologists. The fossil record provides much evidence of the action of evolutionary processes. Conversely, evolutionary models are very helpful in sorting out the historical record. Here, as in physics, the interaction between theory and history depends on the assumption that the underlying processes have remained constant. It is only our knowledge of the initial conditions that is at issue. Biology can tell us little about the origin of life, which now seems primarily a problem for chemistry.

Economics provides the best example from the social sciences. And here the relationship between current economic theory and economic history is less clear than in either physics or biology. Current economic models can at most handle fairly limited contexts. And they are geared to current economic conditions. The development of models applicable to earlier economic conditions, say England in the eighteenth century, is left to economic historians. In general, economics lacks satisfactory models of economic development for which history might supply missing initial conditions and evidence. Not that there have been no attempts at developing such models—remember Marx. But models of large-scale development are clearly the weakest of existing economic models. So the relationship between economic theory and economic history remains weak and fragmented.

The current relationship between theories of science and the history of science cannot be any stronger than the comparable relationship in economics. Existing economic models are far better developed than any current models of science, even those which borrow economic concepts. Testing economic models, whether of current or historical developments, is thus more feasible. On the other hand, just because scientific activities are far more restricted than economic activities, the prospects for a genuine theory of scientific development may be better than those for a theory of economic development. This hope, however, requires that the development of science not be too dependent on general economic development—something many recent sociologists and social historians of science would deny.

My own model of science provides hope for a relationship between history and theory a little more like that in biology. The activities of model construction and model choice abstract from the scientific context in much the same way as models of mutation and selection abstract from the biological environment. These activities may take place in many different social and economic settings.

On the other hand, the conditions under which science has developed have changed more, and more rapidly, than those in biological evolution. This enhances the role of information about the particular circumstances in attempts to apply the general model to particular cases. Much more of what one can say about particular scientific episodes may be simply historical. So, as in economics, we may end up with fairly strong models of science in particular places and times, but weak models of longer-range historical development. This would mean that studies of science in earlier centuries could have only limited relevance to studies of twentieth-century science—contrary to the assumption of many post-positivist historians and philosophers of science who, following Kuhn, move all too easily from the Copernican to the Einsteinian revolution.

Only time will tell whether the biological or the economic analogy is the more appropriate to the theory of science. The *nature* of the relationship, however, is clear. It is the standard relationship between theoretical models and empirical data.

Notes

*The support of the National Science Foundation through its program of research grants in History and Philosophy of Science is hereby gratefully acknowledged. I would also like to thank the Center for Philosophy of Science (University of Pittsburgh) and The Institute for Advanced Study (Princeton) for support and hospitality.

1. The philosophical tenor of the times may be ascertained from the collections of papers edited by Ayer (1959) or Feigl and Maxwell (1961).

2. The recent origins of "cognitive" or "interpretive" sociology of science go back to Barnes (1974) and Bloor (1976). For an overview of what has transpired since then, a good collection of essays, and an extensive bibliography, see Barnes and Edge (1982). Another recent collection with a somewhat different perspective is Knorr-Cetina and Mulkay (1983).

3. See "General Introduction", Tweney, Doherty and Mynatt (1981).

4. Quine (1969).

5. Kuhn (1962, pp. 8–9).

6. I formulated a version of the circle argument (though not by that name) in Giere (1973). I was not attempting to provide an original argument, but merely to give voice to the convictions of middle-of-the-road philosophers of science at the time. This present paper contains my own response to the argument.

7. Recall that Popper's *Logic of Scientific Discovery* appeared in 1959 and the second edition of Carnap's *Logical Foundations of Probability* in 1962.

8. Quine (1969).

9. Lakatos (1970), reprinted in Lakatos (1978).

10. Lakatos (1971), reprinted in Lakatos (1978, p. 132).

11. Laudan (1977, ch. 5) invokes a methodological principle very similar to Lakatos's meta-methodological rule. If both methodologies agreed with our preanalytic judgments about established cases of good and bad science, as is surely possible, what should we do then?

12. I have adapted this line from George Gamow who in turn adapted it from a famous line of Lewis Carroll.

13. Here I have in mind particularly Toulmin (1972) and Campbell (1974). Other naturalistic epistemologists, such as Shimony (1981), do not make such direct use of the evolutionary analogy.

14. The distinction between an exemplar and a disciplinary matrix was made in Kuhn (1974), reprinted in Kuhn (1977).

15. I have developed the general ideas of this section at somewhat greater length in a recent paper (1984) bearing the same title. The reference is to the "definitional" (also called "semantic") view of theories developed by van Fraassen (1970), (1980), and Suppe (1973), and also by Suppes (1967), Sneed (1971), and Stegmueller (1976; 1979). My use of these notions, as described in Giere (1979; 2d ed., 1984) is much less normative than that of these authors, especially the latter three. In particular, I see no special virtue in requiring a set-theoretical representation of theories. In many ways my views are closer to those of Wartofsky (1979) who takes models as a type of representation.

16. For recent statements of the argument by sociologists see Barnes (1982), Bloor (1982), or Barnes and Bloor (1982). For Mary Hesse's presentation, see chapters 1 and 2 of her (1974). An early version of the argument may also be found in Kuhn (1961), reprinted in Kuhn (1977).

17. See Berlin and Kay (1969). Even Bloor (1982, p. 279) seems willing to admit the force of biology in some perceptual judgments.

18. Among recent historians and philosophers who have written on the revolution in geology are Frankel (1979; 1982), R. Laudan (1981), and Ruse (1981).

19. See, for example, the bibliographies to the references cited in n. 18.

20. See Simon (1945; 1957; 1959; and 1972).

21. See Kuhn (1961).

22. This is a too cryptic, but nevertheless appropriate, description of the view developed in Latour and Woolgar (1979).

23. Edge and Mulkay (1976).

24. Among post-positivist philosophers of science, Lakatos places most stress on the importance of success.

References

Ayer, A. J., ed. 1959. *Logical Positivism.* Glencoe, Ill.: The Free Press.

Barnes, Barry. 1974. *Scientific Knowledge and Sociological Theory.* London: Routledge & Kegan Paul.

———. 1982. *T. S. Kuhn and Social Science.* London: Macmillan.

Barnes, Barry, and Bloor, David. 1982. "Relativism, Rationalism and the Sociology of Knowledge", in M. Hollis and S. Lukes, eds., *Rationality and Relativism,* pp. 21–47. Cambridge, Mass.: MIT Press.

Barnes, Barry, and Edge, David, eds. 1982. *Science in Context.* Cambridge, Mass.: MIT Press.

Berlin, B., and Kay, P. 1969. *Basic Color Terms: Their Universality and Evolution.* Berkeley: University of California Press.

Bloor, David. 1976. *Knowledge and Social Imagery.* London: Routledge & Kegan Paul.

———. 1982. "Durkheim and Mauss Revisited: Classification and the Sociology of Knowledge". *Studies in the History and Philosophy of Science,* pp. 267–297.

Campbell, Donald T. 1974. "Evolutionary Epistemology", in P. A. Schilpp, ed., *The Philosophy of Karl Popper,* pp. 413–463. La Salle, Ill.: Open Court.

Carnap, Rudolf. 1962. *Logical Foundations of Probability.* 2d ed. Chicago: University of Chicago Press.

Edge, David, and Mulkay, Michael. 1976. *Astronomy Transformed: The Emergence of Radio Astronomy in Britain.* New York: Wiley.

Feigl, H., and Maxwell, G., eds. 1961. *Current Issues in the Philosophy of Science.* New York: Holt, Rinehart & Winston.

Frankel, Henry. 1979. "The Career of Continental Drift Theory: An Application of Imre Lakatos' Analysis of Scientific Growth to the Rise of Drift Theory". *Studies in the History and Philosophy of Science* 10: 305–324.

———. 1982. "The Development, Reception and Acceptance of the Vine-Matthews-Morley Hypothesis". *Historical Studies in the Physical Sciences* 13: 1–39.

Giere, Ronald N. 1973. "History and Philosophy of Science: Intimate Relationship or Marriage of Convenience?". *British Journal for the Philosophy of Science* 24: 282–297.

———. 1984. *Understanding Scientific Reasoning.* 2d ed. New York: Holt, Rinehart & Winston.

———. 1984. "Constructive Realism", in P. M. Churchland and C. Hooker, eds., *Images of Science.* Chicago: University of Chicago Press.

Hesse, Mary B. 1974. *The Structure of Scientific Inference.* Berkeley: University of California Press.

Knorr-Cetina, K. D., and Mulkay, Michael, eds. 1983. *Science Observed.* Hollywood: Sage Publications.

Kuhn, T. S. 1961. "The Function of Measurement in Modern Physical Science". *Isis:* 161–190.

———. 1962. *The Structure of Scientific Revolutions.* Chicago: University of Chicago Press.

———. 1974. "Second Thoughts on Paradigms", in F. Suppe, ed., *The Structure of Scientific Theories.* Urbana, Ill.: University of Illinois Press.

———. 1977. *The Essential Tension.* Chicago: University of Chicago Press.

Laudan, Larry. 1977. *Progress and Its Problems.* Berkeley: University of California Press.

Laudan, Rachel. 1981. "The Recent Revolution in Geology and Kuhn's Theory of Scientific Change", in P. D. Asquith and Ian Hacking, eds., *PSA, 1978* 2: 227–239. East Lansing, Mich.: The Philosophy of Science Association.

Lakatos, I. 1970. "Falsification and the Methodology of Scientific Research Programmes", in

I. Lakatos and A. Musgrave, eds., *Criticism and the Growth of Knowledge*. Cambridge: Cambridge University Press.

———. 1971. "History of Science and Its Rational Reconstruction", in *PSA, 1970*. Boston Studies in the Philosophy of Science 8: 91–135. Dordrecht: D. Reidel.

———. 1978. *Philosophical Papers*. Vol. I, John Warrall and Gregory Currie, eds. Cambridge: Cambridge University Press.

Latour, Bruno, and Woolgar, Steve. 1979. *Laboratory Life*. Beverly Hills: Sage Publications.

Popper, K. R. 1959. *The Logic of Scientific Discovery*. London: Hutchinson.

Quine, W. V. O. 1969. "Epistemology Naturalized", in *Ontological Relativity and Other Essays*. New York: Columbia University Press.

Ruse, Michael. 1981. "What Kind of a Revolution Occurred in Geology?", in P. D. Asquith and Ian Hacking, eds., *PSA, 1978* 2: 240–273. East Lansing, Mich.: The Philosophy of Science Association.

Shimony, Abner. 1981. "Integral Epistemology", in Marilynn B. Brewer and Barry E. Collins, eds., *Scientific Inquiry and the Social Sciences*, pp. 98–123. San Francisco: Jossey-Bass.

Simon, H. A. 1945. *Administrative Behavior*. New York: Macmillan.

———. 1957. *Models of Man, Social and Rational*. New York: Wiley.

———. 1959. "Theories of Decision-Making in Economics and Behavioral Science", *American Economic Review*: 235–283.

———. 1972. "Theories of Bounded Rationality", in Roy Radner and C. B. McGuire, eds., *Decision and Organization*, pp. 161–176. Amsterdam: North-Holland.

Sneed, J. D. 1971. *The Logical Structure of Mathematical Physics*. Dordrecht: D. Reidel.

Stegmueller, W. 1976. *The Structure and Dynamics of Theories*. New York: Springer.

———. 1979. *The Structuralist View of Theories*. New York: Springer.

Suppe, F. 1973. "Theories, Their Formulations, and the Operational Imperative". *Synthese* 25: 129–164.

Suppes, P. 1967. "What Is a Scientific Theory?", in S. Morgenbesser, ed., *Philosophy of Science Today*. New York: Basic Books.

Toulmin, S. 1972. *Human Knowledge*. Princeton: Princeton University Press.

Tweney, Ryan D., Doherty, Michael E., and Mynatt, Clifford R., eds. 1981. *On Scientific Thinking*. New York: Columbia University Press.

van Fraassen, B. C. 1970. "On the Extension of Beth's Semantics of Physical Theories", *Philosophy of Science* 37: 325–339.

———. 1980. *The Scientific Image*. Oxford: Oxford University Press.

Wartofsky, Marx W. 1979. *Models*. Dordrecht: D. Reidel.

The Philosopher of Science
as Expert Witness

Philip L. Quinn

When I was invited to contribute to this volume, the editors expressed the hope that the essays in it would paint "an overall picture of where philosophy of science is going in the near future". I hesitated. Since I am no prophet, could there be any profit in my speculating on the future of philosophy of science? In the course of the past generation or so, philosophy of science in the anglophone world has achieved professional autonomy. It has its own specialist journals and professional meetings; independent programs and departments in history and philosophy of science have taken root and flourished in major universities. Mastery of a formidable body of technical background knowledge has become a prerequisite for speaking with authority about a wide range of problems in philosophy of physics, philosophy of biology, philosophy of psychology, and philosophy of economics. Ties with other, more traditional areas of philosophy have been loosened if not dissolved; new jargon has sprung up like weeds. In short, philosophy of science has acquired a turf of its own to defend, become inaccessible to outsiders, and taken on the trappings of a typical academic specialization. So it would seem that the agenda for philosophy of science in the near future will be set by the insiders in each of its subspecializations. Whatever problems they find tractable will dominate discussion in the journals and at meetings. And since tractability is relative to the tools that happen to be at hand, unpredictable opportunism would appear to be the preferred tactic for philosophers of science, as it has often proved to be for scientists.

Yet this can hardly be the whole story. Philosophy of science is not so hermetically sealed off, not yet anyway, from the broader currents of contemporary intellectual life that it is completely insensitive to influences and pressures from the outside. Like other areas of philosophy, philosophy of science has not yet become nothing but a glass bead game. Indeed, external pressure may be producing movement in quite the opposite direction. Contemporary philosophy, so we are told, has taken an applied turn during the past twenty years. Starting with a renewal of interest in normative questions in ethics and political philosophy in the late sixties, the process has been gathering momentum ever since. Now there are philosophical societies, journals, programs or meetings devoted to medical ethics, legal ethics, business ethics, engineering ethics, environmental ethics, ethics and animals, and agricultural ethics; and the end of the process is nowhere in sight. It is, of course, no accident that the process began at a time when the United States was bogged down in a despicable war in Southeast Asia. Nor is it entirely coincidental that it has continued during a period of hard times in the academic marketplace. Philosophers are no less prone than others to looking for a justification of their activities in terms of social utility and for a match between their intellectual interests and economic demand. What was good enough for the scientific community in the forties, when Vannevar Bush wrote his ideological masterpiece *Science: The Endless Frontier,* was not to be despised by the philosophical community of the seventies and eighties. And so the word went out that philosophers too had their bit of expertise to contribute to the formulation and implementation of public policy. Could philosophers of science join the ethicists in taking the applied turn? Should they?

At first glance, it would appear that philosophers of science were ideally positioned to jump on the bandwagon. We live, we are told, in an age of science. The present division of intellectual labor confers unprecedented cognitive authority on what is taken to be reputable scientific thinking. It is a mark of enlightened lay opinion to defer to what the scientific experts say on an immense variety of topics. If we are told that it is scientifically respectable to believe that the universe originated in a big bang billions of years ago, we nod assent without further ado, even if we have a very imperfect grasp of what is actually being asserted and no idea at all about what evidence could be adduced to support such an assertion. If we are told that it is not scientifically respectable to believe that ancient astronauts had an airfield constructed on the plain of Nazca, we hasten to agree, even if the hypothesis of ancient astronauts seems intuitively plausible. And if we are told that respectable scientific opinion is divided about whether there is a safe level of radiation exposure, we think it rationally incumbent upon ourselves to try not to panic immediately if we learn we have been exposed to low-level radiation. But whence comes the authority of scientific opinion? And how is the boundary around respectable scientific opinion to be drawn? Most scientists would be hard pressed to give decent answers to such questions. Like the practitioners of other crafts, scientists tend to be too busy doing science to spend much

time reflecting upon or articulating the epistemological presuppositions of
their characteristic activities. Textbook clichés about the scientific method, im-
perfectly remembered, serve as proxies for serious thought about such questions in the minds of many scientists. By contrast, philosophers of science are
supposed to be experts on such second-order questions. From historical case
studies, they are supposed to have learned what has differentiated scientific
thinking from superstition and other modes of thought. From rational reconstructions of the achievements of contemporary science, they are supposed
to have derived insight into the grounds of science's epistemic authority. So
philosophers of science ought to be able to provide politicians and granting
agencies with good advice about which proposals have the earmarks of respectable science and deserve funding, and they ought to be able to explain
to the educated lay public even better than most scientists why confidence
in respectable scientific opinion is justified and what counts as respectable scientific opinion. Thus it would seem that philosophers of science are well placed
to take the applied turn and to grab a piece of the policy-making action.

   As a matter of fact, philosophers of science have been rather slow to
inject themselves into the process of creating public policy. By way of illustration, consider the proceedings of a conference held in 1977, sponsored by the
Philosophy of Science Association and funded by the National Science Foundation, on critical research problems in philosophy of science.[1] These proceedings contain papers on methodological approaches to philosophy of science,
philosophy of science and other disciplines, meta-science topics, and philosophical foundations of various scientific disciplines. Apart from some sensible but
rather mundane suggestions in one paper about incorporating some results
from philosophy of science into science education, they contain no discussion
of whether philosophy of science has any bearing on what goes on outside
the groves of academe. Only within the past two years have we seen a case
in which philosophy of science has played a major and well-documented role
in a significant policy decision. This case has all the makings of an interesting
precedent. It has already generated polite but fierce controversy among philosophers of science both in print and in discussion and is likely to provoke
further debate. It illustrates vividly some of the pitfalls which await philosophers of science who apply their expertise to issues in the political arena. So
if we subject this case to close scrutiny, we may be able to arrive at some tentative conclusions about whether an applied turn in philosophy of science is
likely to lead into the corridors of power or only into a dead-end street. And
these conclusions may help us make up our minds about whether we think
it a good idea to respond to whatever pressures there may be toward an applied

turn in philosophy of science by giving in to them or by resisting them. We shall then be in a better position to address the question of whether this is a direction in which philosophy of science ought to go in the near future.

The case I have in mind is, of course, the notorious legal case *McLean v. Arkansas*. On March 19, 1981, the Governor of Arkansas signed into law Act 590 of 1981, entitled the "Balanced Treatment of Creation-Science and Evolution-Science Act", which required balanced treatment of creation science and evolutionary biology in the public schools of Arkansas. On May 27, 1981, a group of plaintiffs mounted a challenge to the Act's constitutional validity by bringing suit in U.S. District Court against the Arkansas Board of Education. The plaintiffs alleged that Act 590 violated the Establishment Clause of the First Amendment. Trial began on December 7, 1981. The Court's Opinion, handed down by Judge William R. Overton on January 5, 1982, held the Act to be unconstitutional. Michael Ruse, a prominent philosopher of biology, provided a number of position papers for the plaintiffs, and testified as an expert witness for the plaintiffs, on the defining characteristics of science. Judge Overton's Opinion acknowledges that its statement of the essential characteristics of science is based in large part on Ruse's testimony. On the basis of this statement of the essential characteristics of science, the Opinion argues that creation science is not real science. Subsequent to the Court's decision, Larry Laudan, another prominent philosopher of science, severely criticized the argumentation of the Opinion. Ruse responded with a spirited defense of Judge Overton's reasoning and, incidentally, of his own views on the nature of science. Later Laudan argued that Ruse had not successfully replied to his original criticisms.

In this paper, I propose to do a detailed analysis of this controversy. In the course of the discussion it will become clear that I agree with many but not all of Laudan's critical points. But I think more needs to be said in criticism both of Judge Overton's Opinion and of Ruse's philosophical views. First, I shall lay out the main lines of the relevant portion of the argument in Judge Overton's Opinion and try to show that a crucial part of the argument is unsound. Then I shall consider the question of whether the argument of the Opinion accurately reflects the views Ruse has expressed in print. I shall argue that, although judging by what Ruse says, there is some ambiguity about what views he actually holds, the Opinion is in essential respects a good representation of his published views. From these arguments, I shall extract some lessons about the dangers philosophers of science let themselves in for when they assume the role of expert witness. Because I know Ernan McMullin has a long-standing interest in both the methodological questions I shall be discussing and the problem of the proper relation between theology and science raised by *McLean* v. *Arkansas,* I hope he will find the story I have to tell engaging and, perhaps, somewhat amusing.

Judge Overton's Opinion

According to the Opinion, a statute violates the Establishment Clause if it fails any part of the following three-pronged test:

> First, the statute must have a secular legislative purpose; second, its principal or primary effect must be one that neither advances nor inhibits religion . . . ; finally, the statute must not foster "an excessive government entanglement with religion".[2]

Though a proof that Act 590 failed on any one part of this three-pronged test would have been sufficient to show that it was unconstitutional, the Opinion argues that the Act fails on all three parts. Since the issues of importance from the point of view of philosophy of science all arise in the course of the argument that the Act fails the second part of the test, I shall confine my analysis to the Opinion's reasoning to the conclusion that the Act's primary effect is the advancement of religion. Because I am going to be very critical of that reasoning, I should say at the outset that I find the argument for the conclusion that Act 590 fails the first part of the test entirely convincing. The Opinion cites precedents to show that courts are not bound to consider only legislative statements of purpose or legislative disclaimers in determining the legislative purpose of a statute but may also consider such factors as the historical context of a statute, the specific sequence of events leading up to passage of the statute, departures from normal procedural sequences, substantive departures from the normal, and contemporaneous statements of the legislative sponsor. Appealing to evidence which is relevant to such contextual factors, the Opinion argues that Act 590 was passed by the Arkansas General Assembly with the specific purpose of advancing religion and so fails to have a secular legislative purpose. I agree. And since failure on this part of the three-pronged test is, as noted above, by itself sufficient to show the Act to be in violation of the Establishment Clause, I also concur with Judge Overton's ultimate conclusion that Act 590 is unconstitutional. If the Opinion had argued no more than this, which would have been enough to dispose of Act 590, I would have no quarrel with it. Unfortunately, the Opinion showed no such admirable restraint.

Judge Overton's argument to the conclusion that Act 590 has the advancement of religion as its primary effect may be schematically reconstructed as follows:

(1) Act 590 does have the advancement of religion as a major effect.
(2) Act 590 does not have the advancement of science as an effect.
(3) Act 590 does not have the advancement of any other thing as an effect.
(4) Hence, Act 590 has the advancement of religion as its only effect.
(5) Whence, Act 590 has the advancement of religion as its primary effect.

The argument is intuitively valid: (4) follows from (1)–(3), and (5) follows from (4). But is it sound? It would appear that each of (1)–(3) needs further justification.

It is curious that Judge Overton says nothing at all to justify (3). Indeed, the Opinion shows no indication that he is aware of the need to assume (3), or something very much like it, to insure the validity of its argument. Speaking to this point, Laudan takes Judge Overton to task for making "the claim, specious in its own right, that since Creationism is not 'science' it must be religion".[3] In response, Ruse tries to defend Judge Overton against the charge of having adopted "the naive dichotomy of 'science or religion but nothing else'".[4] But Judge Overton does not make the specious claim Laudan attributes to him, and Ruse's reply does nothing to show that Judge Overton is not committed to a somewhat different but equally dubious thesis. Both Laudan and Ruse have missed the point; neither has seen where the real problem with this part of Judge Overton's argument lies.

Judge Overton first argues that creation science is, or at least is inspired by, religion. This argument supports (1), and I will have more to say about it later on. He then gives an independent argument to show that creation science is not science. This argument supports (2), and I will later analyze it in great detail. Finally, Judge Overton couples the conclusions of the previous arguments in this passage:

> The conclusion that creation science has no scientific merit or educational value as science has legal significance in the light of the Court's previous conclusion that creation science has, as one major effect, the advancement of religion. The second part of the three-pronged test for establishment reaches only those statutes having as their *primary* effect the advancement of religion. Secondary effects which advance religion are not constitutionally fatal. Since creation science is not science, the conclusion is inescapable that the *only* real effect of Act 590 is the advancement of religion.[5]

But the conclusion is very far indeed from being inescapable. Though it does not rest on the specious claim that since creation science is not science it must be religion, it does rest on the claim that since creation science advances religion but not science it advances only religion. This claim is certainly not obvious and needs to be supported by argument. To argue for it would be to argue in support of something like (3) which would rule out other possibilities. This the Opinion does not do, and the failure to do it is a serious flaw in the Opinion. As long as the possibility that creation science is, above and beyond being religion and not science, some other thing as well (perhaps, for example, speculative philosophy) has not been ruled out, it remains an open question whether Act 590 also has the advancement of some such other thing as a major effect, and hence it remains an open question whether Act 590 has the advancement of this other thing as its primary effect and the advancement of religion only as its secondary effect. Since, as Judge Overton says, secondary effects which advance religion are not constitutionally fatal, such questions cannot be left open. The Opinion, however, contains no arguments for a principle like (3) which would close them. Though, for all I know, (3) may be true and so, for all I have said so far, the argument in which it is a premise may be sound, that argument is unconvincing because (3) is not obviously true and the Opinion furnishes no evidence to support it. Perhaps the best that can be done by way of a defense of Judge Overton's reasoning in the passage quoted above is to suppose that, since the defendents had not claimed that Act 590 has the advancement of something other than science as an effect, Judge Overton thought there was no need for him to consider possibilities that had not come up in the course of the trial. If Judge Overton did think something like this, I would say he made a mistake, for his failure to rule out such possibilities makes his argument less than rationally compelling.

As I mentioned before, the Opinion does contain an argument for (1). It begins with a citation from Section 4 of Act 590:

> *Definitions.* As used in this Act:
> (a) "Creation-science" means the scientific evidences for creation and inferences from those scientific evidences. Creation-science includes the scientific evidences and related inferences that indicate: (1) Sudden creation of the universe, energy, and life from nothing; (2) The insufficiency of mutation and natural selection in bringing about development of all living kinds from a single organism; (3) Changes only within fixed limits of originally created kinds of plants and animals; (4) Separate ancestry for man and apes; (5) Explanation of the earth's geology by catastrophism, including the occurrence of a world-wide flood; and (6) A relatively recent inception of the earth and living kinds.[6]

According to the Opinion, "Section 4(a) is unquestionably a statement of religion, with the exception of 4(a)(2) which is a negative thrust aimed at what the creationists understand to be the theory of evolution."[7] As I understand Judge Overton's reasoning, he deploys two distinct arguments in support of this claim. The more general argument aims to establish that the reference to creation from nothing in 4(a)(1) is religious. To show that the concept of creation from nothing is the concept of creation by God, Judge Overton appeals to uniformity of testimony. He notes, first, that all the theologians who testified, including defense witnesses, were of the opinion that the concept of creation from nothing is the concept of creation by God and, second, that leading creationist writers are of the same opinion. To show that the concept of creation by God is religious, Judge Overton appeals to legal precedent. He cites from *Malnak* v. *Yogi* the opinion that concepts concerning a supreme being are religious and do not cease to be so in virtue of being presented as philosophy or science. Since (i) the concept of creation from nothing is the concept of creation by God, and (ii) the concept of creation by God is the concept of creation by a supreme being, and (iii) any concept concerning a supreme being is a religious concept, the reference to creation from nothing in 4(a)(1) is therefore religious. The more specific argument aims to show both that Section 4(a) as a whole is congruent with the literal interpretation of Genesis favored by fundamentalist Christians and that the ideas in 4(a)(1) in particular are identical to the ideas in the Genesis creation story as literally interpreted. To support the congruence claim Judge Overton enumerates extensive and specific parallels between Section 4(a) and Genesis. They include: (i) the parallel between 4(a)(1) and Genesis 1; (ii) the parallel between 4(a)(5) and Genesis 7 and 8; and (iii) the parallel between the use of the term 'kinds' in 4(a)(2), 4(a)(3) and 4(a)(6) and its use in Genesis. To support the identity claim Judge Overton says only that the ideas of 4(a)(1) are parallel to the creation story in Genesis 1 and are not parallel to any other creation story. From the combination of the two arguments, Judge Overton draws the conclusion that there is "no doubt that a major effect of the Act is the advancement of particular religious beliefs".[8]

Because my purpose in this paper is not to deal in depth with the tangled knot of issues in theology and scriptural exegesis raised by this part of the Opinion, I have taken the liberty of stating Judge Overton's arguments briefly and informally. Even so, some of his claims should not go unchallenged. Three points deserve special emphasis. First, though there may be a technical, legal sense in which any use of the concept of a supreme being counts as religious, this is not a sense in which any such use advances or promotes religion.

After all, that concept is used in purely theoretical discussions in works of metaphysics that have no tendency to promote or advance religious belief or practice by any individual or group. So there is plenty of room to doubt that Judge Overton's more general argument does anything at all to support the conclusion quoted above. Second, despite the assertions of some fundamentalist Christians, it is unlikely that the best literal interpretation of the creation story in Genesis 1 makes reference to creation from nothing. The authors of that creation story seem to think of God's creative activity as consisting of imposing form and structure on a pre-existent formless stuff, which scholars identify with the primordial ocean of ancient Semitic cosmogonies.[9] And so there is also plenty of room to doubt Judge Overton's claim that the ideas of 4(a)(1) are identical, and not merely similar, to the ideas in the Genesis creation story as literally interpreted. Third, though there are extensive parallels between Section 4(a) and Genesis, it is doubtful that this by itself is sufficient to show that a major effect of Act 590 is to advance particular religious beliefs. After all, there are also extensive parallels between parts of the criminal law and Deuteronomy. Both the Decalogue in Deuteronomy 5 and the criminal law prohibit such things as murder, theft and perjury. But surely this does not suffice to show that a major effect of the criminal law is to advance particular religious beliefs. Hence, I think there are solid rational grounds for doubting that Judge Overton's argument for (1) is successful.

As I have said before, the Opinion also contains an argument for (2). Because this is the argument to which philosophy of science is supposed to make some contribution, I am going to analyze it in detail. Judge Overton begins with a statement of what he takes to be the essential characteristics of science:

(1) It is guided by natural law;
(2) It has to be explanatory by reference to natural law;
(3) It is testable against the empirical world;
(4) Its conclusions are tentative, i.e., are not necessarily the final word; and
(5) It is falsifiable (Testimony of Ruse and other science witnesses).[10]

The Opinion does not make it clear whether Judge Overton takes these characteristics to be conditions both necessary and sufficient for scientific status. However, the argument Judge Overton bases on them requires, if it is to be valid, that they be necessary conditions, but does not require that they also be sufficient conditions, for scientific status. So I shall proceed to reconstruct the argument by attributing to Judge Overton *only* the claim that the characteristics in question are necessary conditions for scientific status. Since Judge Overton is trying to show that creation science is not science but not that evolution science is science, he does not need to take a stand on the issue of which conditions are sufficient for scientific status.

The strategy of the Opinion is to argue that each of the positive theses of Section 4(a) of Act 590 lacks one or more of the relevant characteristics and is for that reason unscientific. With respect to 4(a)(1), Judge Overton claims that the hypothesis of creation from nothing "is not science because it depends upon a supernatural intervention which is not guided by natural law. It is not explanatory by reference to natural law, is not testable, and is not falsifiable."[11] Section 4(a)(2) is dismissed as "an incomplete negative generalization directed at the theory of evolution".[12] With reference to 4(a)(3), Judge Overton complains that it "fails to conform to the essential characteristics of science for several reasons. First, there is no scientific definition of 'kinds' and none of the witnesses was able to point to any scientific authority which recognized the term or knew how many 'kinds' existed. . . . Second, the assertion appears to be an effort to establish outer limits of changes within species. There is no scientific explanation for these limits which is guided by natural law and the limitations, whatever they are, cannot be explained by natural law."[13] About 4(a)(4) Judge Overton says only that "it explains nothing and refers to no scientific fact or theory".[14] According to Judge Overton, 4(a)(5) "completely fails as science". Because the flood referred to in 4(a)(5) is the Noachian flood and even creationist writers concede that it depends upon supernatural intervention, such a "worldwide flood as an explanation of the world's geology is not the product of natural law, nor can its occurrence be explained by natural law".[15] And, as for 4(a)(6), Judge Overton thinks it too "fails to meet the standards of science" because the phrase 'relatively recent inception' has "no scientific meaning"; it can only be given meaning in the context of Act 590 by reference to creationist estimates that the earth is between 6,000 and 20,000 years old, and such a procedure is "not the product of natural law; not explainable by natural law; nor is it tentative".[16] But if each of the positive theses of Section 4(a) is unscientific, then creation science as defined by Act 590 is not science, and hence Act 590 does not have the advancement of science as an effect.

Judge Overton's argument can therefore be adequately represented in the following schematic manner:

(6) If any statement S is scientific, then S either is a natural law or is explainable by a natural law and is testable, tentative and falsifiable.

(7) The statement in 4(a)(1) is neither a natural law nor explainable by a natural law and is neither testable nor falsifiable.

(8) Hence, the statement in 4(a)(1) is not scientific.

(9) The statement in 4(a)(3) is neither a natural law nor explainable by a natural law.

(10) Hence, the statement in 4(a)(3) is not scientific.

(11) The statement in 4(a)(4) is neither a natural law nor explainable by a natural law.

(12) Hence, the statement in 4(a)(4) is not scientific.

(13) The statement in 4(a)(5) is neither a natural law nor explainable by a natural law.

(14) Hence, the statement in 4(a)(5) is not scientific.

(15) The statement in 4(a)(6) is neither a natural law nor explainable by a natural law and is not tentative.

(16) Hence, the statement in 4(a)(6) is not scientific.

(17) Thus, none of the positive statements definitive of creation science in Section 4(a) of Act 590 is scientific.

(18) Whence, creation science as defined by Act 590 is not scientific.

(19) If creation science as defined by Act 590 is not scientific, then Act 590 does not have the advancement of science as an effect.

(20) Therefore, Act 590 does not have the advancement of science as an effect.

I have deliberately constructed this argument in such a way that it is intuitively valid. Unfortunately, it is all too clear that it is unsound. The problem is that (6) is demonstrably false. None of the characteristics it alleges to be necessary conditions for an individual statement to have scientific status is, in fact, a necessary condition of scientific status of an individual statement.

Consider first the condition of either being a natural law or being explainable by a natural law. As Laudan has pointed out, scientists have for a long time understood the difference between establishing the existence of phenomena and explaining them by natural laws.[17] Darwin, for instance, established the existence of natural selection nearly half a century before the discovery of the laws of heredity which help to explain it. If, contrary to fact, there had turned out to be no laws to explain natural selection, Darwin's achievement would still have been scientific. To be sure, as Ruse notes, science looks for explanatory laws.[18] But if there are no laws to be found, scientists are prepared to settle for less and can do so without forfeiting the scientific status of their achievements. Certain statements about individual events in the quantum domain are not laws and have no known explanations in terms of laws; moreover, they can have no explanation in terms of laws if contemporary quantum theory is correct, as it seems to be. But they will remain scientific statements even if contemporary quantum theory is correct. Hence, either being a natural law or being explainable by a natural law is not a necessary condition for scientific status. Thus, the arguments for (10), (12), and (14) fail.

Consider next the conditions of testability and falsifiability. As a result of the work of Pierre Duhem, it has been known to philosophers of science for three-quarters of a century that many scientific statements are neither testable nor falsifiable individually and in isolation but only conjunctively and in corporate bodies. Hence, being testable and being falsifiable are not necessary for individual statements to have scientific status, and the argument for (8) fails too. Moreover, it would not strengthen Judge Overton's argument to retreat to the more plausible claim that only in the case of whole theories, and not on the level of each individual statement, do testability and falsifiability count as necessary conditions for scientific status. Creation science as defined in Section 4(a) of Act 590 and as further interpreted by Judge Overton himself clearly satisfies these conditions. For example, the statements in 4(a)(1) and 4(a)(6), as Judge Overton interprets them, together imply that there is no matter on earth more than 20,000 years old. The trouble with this claim is not that it is untestable or unfalsifiable. Its problem is rather that it has been repeatedly tested and is so highly disconfirmed that, for all practical purposes, it has been falsified.

Unfortunately, the patently false claim that creation science is neither testable nor falsifiable seems well on its way to becoming, for some evolutionary biologists, a rhetorical stick with which to belabor their creationist opponents. In a recent collection of essays, Stephen Jay Gould claims that "'scientific creationism' is a self-contradictory nonsense phrase precisely because it cannot be falsified".[19] And in another essay in the same collection Gould has this to say about creationists:

> They present no testable alternative but fire a volley of rhetorical criticism in the form of unconnected, shaky factual claims — a potpourri (literally, a rotten pot, in this case) of nonsense that beguiles many people because it masquerades in the guise of fact and trades upon the false prestige of supposedly pure observation.[20]

Ironically, in the next sentence Gould goes on to contradict himself by asserting that "the individual claims are easy enough to refute with a bit of research".[21] Indeed, some of them are! But since they are easily refuted by research, they are after all falsifiable and, hence, testable. This glaring incon-sistency is the tip-off to the fact that talk about testability and falsifiabil-ity functions as verbal abuse and not as a serious argument in Gould's anti-creationist polemics.

Finally, consider the condition of tentativeness. Unlike the other condi-tions, tentativeness is not a structural or methodological condition on the con-tent of a body of beliefs but is a psychological condition on the attitudes of believers. But whether a belief is held tentatively or dogmatically is completely irrelevant to whether or not it is scientifically or in any other way epistemi-cally respectable. Laudan puts the point very well:

> Since no law mandates that creationists should be invited into the classroom, it is quite irrelevant whether they themselves are close-minded. The Arkansas statute proposed that Creationism be taught, not that creationists should teach it. What counts is the epistemic status of Creationism, not the cognitive idio-syncrasies of the creationists. Because many of the theses of Creationism are testable, the mind set of creationists has no bearing in law or fact on the merits of Creationism.[22]

Being held tentatively is not a necessary condition for scientific status which the beliefs of creationists fail to satisfy. No matter how dogmatically some contemporary physicists believe the principle of the conservation of mass-energy nothing is done thereby to impugn its scientific status. So the argument for (16) also fails. And since the arguments for (8), (10), (12), (14) and (16) are all unsuccessful, the argument for (17), (18) and (20) also collapses.

Ruse makes two last-ditch attempts to defend Judge Overton's argumen-tative strategy. Neither of them succeeds.

The first involves citing one scientific case in which, so we are to sup-pose, (6) is true and one non-scientific case in which, so we are told, (6) is also true. Ruse says:

> For instance, explanation of the fact that my son has blue eyes, given that both parents have blue eyes, done in terms of dominant and recessive genes and with an appeal to Mendel's first law, is scientific. The Catholic doctrine of tran-substantiation (i.e., that in the Mass the bread and wine turn into the body and blood of Christ) is not scientific.[23]

In the Mendelian example, we are to suppose that each statement is either a natural law or is explainable by a natural law and is testable, falsifiable and tentatively held by the proponent of the explanation. In the other example, we are to suppose that the doctrine of transubstantiation is neither a natural law nor explainable by a natural law and is neither testable, nor falsifiable nor tentatively held by its proponents. But supposing all this gets us exactly no-where. Two positive instances cannot prove a universal generalization which covers other cases, as Ruse himself admits. But a single counterexample can falsify a universal generalization, and between us Laudan and I have provided more than one counterexample to (6).[24] So this line of defense does nothing at all to help Judge Overton's argument.

Ruse's second ploy is to suggest that for legal purposes Judge Overton had to argue that creation science is not science at all because he could not have held Act 590 in violation of the Establishment Clause if he had merely shown that creation science, though testable, has been tested and massively disconfirmed, and is therefore bad or weak science.[25] But this suggestion is mistaken on two counts. First, as I noted above, Judge Overton could have held Act 590 in violation of the Establishment Clause without even addressing the question of the scientific status of creationism merely by arguing, as he in fact did, that Act 590 fails the first part of the three-pronged test. Second, if Judge Overton had been able to show that Act 590 has as a major effect the advancement of religion, then he could at least have tried to argue from the premise that creation science is bad science to the conclusion that Act 590 has the advancement of science only as a minor effect at best. And if he had successfully done this and also shown that Act 590 has no other major effects, then he would have been entitled to conclude that Act 590 has the advance-ment of religion as its primary effect, which is all he needed to establish in order to show that Act 590 fails the second part of the three-pronged test.

There are two conclusions I wish to draw from this portion of my dis-cussion. First, because Judge Overton's Opinion does not provide adequate justification for any of (1)–(3), it is not a rationally persuasive argument for the claim that Act 590 fails the second part of the three-pronged test. Second, because the argument (6)–(20) is demonstrably unsound, philosophy of sci-ence contributed nothing of positive value to the quality of Judge Overton's Opinion. One can only hope that if other judges cite *McLean v. Arkansas* as a precedent, they will have the good sense to ignore the material in Section IV(C).

Ruse's Opinions

It is interesting to ask how we should assign responsibility for the mani-fold errors contained in Judge Overton's argument for (2). Did Judge Overton simply fail to grasp what Ruse and other witnesses told him about how to demarcate science from non-science? Fortunately we are in a position to say something about this question if we examine Ruse's published views on creation science. I consider it fair to proceed in this way because Ruse himself says of his published discussion that it "is the same as what I provided for the plaintiffs in a number of position papers. It also formed the basis of my testimony in court, and, as can be seen from Judge Overton's ruling, was accepted by the court virtually verbatim."[26] As we shall soon see, there appear to be some differences between Ruse's published views and Judge Overton's Opinion. But I shall argue that the real differences are relatively minor and insignificant.

As I pointed out above, Judge Overton must construe the five essential characteristics of science on his list as necessary conditions for scientific status if his argument that any thesis of creation science which lacks one or more of them is for that reason unscientific is to be valid. But Ruse begins by saying: "It is simply not possible to give a neat definition—specifying necessary and sufficient characteristics—which separates all and only those things that have ever been called 'science'."[27] So it might seem at the outset that Judge Overton misinterpreted Ruse when he treated the characteristics on his list as necessary conditions for scientific status. However, things are not quite so simple. Ruse quickly proceeds to muddy the waters. He also says, "Creation-science is not science because there is absolutely no way in which creationists will budge from their position."[28] When push comes to shove and an argument against creationism has to be made, Ruse does treat holding one's views tentatively as a necessary condition for the views held to be scientific. As I argued above, and as Laudan had argued before me, this is a mistake. But when Judge Overton makes it, he displays no misunderstanding of Ruse's real position on creationism. Indeed, it seems he understands it perfectly well.

Ruse's search for "defining features"[29] of science appears, at first glance, to yield a list somewhat different from Judge Overton's. Ruse tells us that "science involves a search for order. More specifically, science looks for unbroken, blind, natural regularities (laws)."[30] He goes on to say that "a major part of the scientific enterprise involves the use of law to effect explanation",[31] including prediction and retrodiction. These are modest enough and sweetly reasonable statements. One tells us only that science looks for laws; it offers no guarantee that science can always succeed in finding them. The other tells us only that a major part of the scientific enterprise as a whole is explanation in terms of law. Neither says or implies that every scientific statement either is a natural law or is explainable by a natural law. But, again, when the crunch comes, Ruse abandons modesty and sweet reason. In his response to Laudan, when he tries to deal with the fact that Charles Lyell in his *Principles of Geology* explicitly left open the possibility that divine intervention is required to account for human origins, Ruse pushes a harder line:

> Science, like most human cultural phenomena, has evolved. What was allowable in the early nineteenth century is not necessarily allowable in the late twentieth century. Specifically, science today does not break with law. And this is what counts for us. We want criteria of science for today, not for yesterday.[32]

So Ruse's concession that there are no neat necessary and sufficient conditions for all and only the things that have ever been called 'science' cuts no ice in practice. All that really matters when Ruse wants a stick with which to beat the creationists is that in the late twentieth-century science allows, so Ruse alleges, no break with law. Hence, once again, Judge Overton displays no real misunderstanding of Ruse's position when he lays it down for purposes of a decision in 1982 that every scientific statement either is a natural law or is explainable by a natural law. And again, as I argued above, the view in question is seriously in error.

To these characteristics, Ruse adds testability, which encompasses both confirmability and falsifiability. Concerning testability in general, he says that "a genuine scientific theory lays itself open to check against the real world".[33] And with respect to falsifiability, he says that "a body of science must be *falsifiable*".[34] Hence, although Ruse is not committed, as Judge Overton is, to the false claim that each individual scientific statement is testable and falsifiable in isolation, he is committed to the view that testability and falsifiability are necessary conditions of scientific status for theories as wholes. But, as I pointed out before, creation science as defined in Section 4(a) of Act 590 satisfies these conditions. Ruse disagrees. He asserts that "creation scientists do little or nothing by way of genuine test".[35] Even if true, this remark is completely irrelevant. The requirement is that a scientific theory be testable, not that its proponents actually test it. However, Ruse goes on to say, even if creation scientists do expose their theories to tests, "when new counter-empirical evidence is discovered, creation scientists appear to pull back, refusing to allow their position to be falsified".[36] This remark too, even if accurate, is utterly beside the point. The requirement is that a theory *be* falsifiable by empirical evidence, not that its adherents admit that it has been falsified if and when it has been. Once creation scientists make testable assertions, as they have, it is up to the evidence and not to them whether those assertions are disconfirmed to the point of being falsified. Hence, Ruse's main reasons for considering creation science untestable and unfalsifiable turn out to be, upon inspection, nothing more than two irrelevant *ad hominem* arguments.

My discussion of Ruse's position thus far yields two conclusions. First, though some of Ruse's statements are more cautious and nuanced than parallel statements by Judge Overton, when it comes time to mount an argument against creation science both of them are committed to the untenable view that being either a natural law or explainable by a natural law and being held tentatively by its adherents are conditions necessary for a statement to be scientific. Second, although Ruse does not make the mistake Judge Overton does in assuming that the criteria of testability and falsifiability apply to all individual statements of science, this is of no help in constructing a case against the scientific status of creation science because creation science as defined in Act 590 satisfies both these criteria. Both Ruse's arguments against this conclusion are irrelevant because fallaciously *ad hominem*.

There is one characteristic on Ruse's list of the defining features of science that is not precisely reflected anywhere on Judge Overton's list. This is what Ruse thinks of as professional integrity. Describing this characteristic, Ruse says:

> A scientist should not cheat or falsify data or quote out of context or do any other thing that is intellectually dishonest. Of course, as always, some individuals fail; but science as a whole disapproves of such actions. Indeed, when transgressors are detected, they are usually expelled from the community.[37]

In the light of what has already been said, it should be tolerably clear that integrity, like tentativeness, is a characteristic of persons either individually or as groups. As such, it has no bearing on the scientific status of their ideas or assertions. I have no wish to defend, mitigate or excuse the intellectual dishonesty one finds in the writings of some creationists. But I do detect a certain unctuous tone in Ruse's praise of the intellectual honesty of scientists in general at a time when scientific fraud has become something of a national scandal in the United States. All over the country institutions are busy putting in place formal and explicit policies for dealing with dishonesty in research, and clearly the impetus to do so comes mainly from the numerous and well-publicized cases of scientific fraud that have occurred in recent years. The fact that institutions find it necessary to make formal policy on such matters is a good indication that the informal self-policing mechanisms of the scientific community are not doing their job.[38]

We are now in a position to answer the question asked at the beginning of this section. Judge Overton did not seriously misunderstand Ruse's position. Although some of Ruse's views are carefully qualified when abstractly stated, the qualifications are omitted when it comes time to argue that creation science is not real science, and the results are philosophical mistakes which match in most important respects the errors in Judge Overton's Opinion. And even where there are differences between their views, as in the case of the unit to which the criteria of testability and falsifiability are properly to be applied, the only result is that they reach similar false conclusions by means of somewhat different bad arguments.

In a way, this is a pity. By allowing the issue to turn entirely on whether creation science is or is not real science, Judge Overton missed other opportunities to argue that Act 590 fails the second part of the three-pronged test. A remark by Laudan will help us focus our attention on one such possibility:

> Rather than taking on the creationists obliquely and in wholesale fashion by suggesting that what they are doing is "unscientific" *tout court* (which is doubly silly because few authors can even agree on what makes an activity scientific), we should confront their claims directly and in piecemeal fashion by asking what evidence and arguments can be marshalled for and against each of them. The core issue is not whether Creationism satisfies some undemanding and highly controversial definitions of what is scientific; the real question is whether the existing evidence provides stronger arguments for evolutionary theory than for Creationism.[39]

The question is not whether creation science fails to accord with some dubious and probably ephemeral theories about what is necessary for counting as science. The real issue is whether creation science, whatever it may be, now has high epistemic status as compared to its rivals for credibility in the empirical domain. Since it does not, the following argument seems promising;

(21) Act 590 does have the advancement of religion as a major effect.

(22) Act 590 does not have the advancement of empirical knowledge as a major effect.

(23) Act 590 does not have the advancement of any other aim as a major effect.

(24) Hence, Act 590 has the advancement of religion as its only major effect.

(25) Whence, Act 590 has the advancement of religion as its primary effect.

Assuming that a better argument for (21) than the one Judge Overton gave could be found, the arguments for (22) and (23) would be relatively straightforward. In outline, here is the argument for (22). Because creation science does make empirically testable claims, it is appropriate to judge its epistemic status by how well those claims have stood up under testing. But those of the empirical claims made by creation science which have been subjected to testing have not fared well at all; they have failed to be confirmed and many have been highly disconfirmed. Hence, creation science contributes at best indirectly and in a minor way to our empirical knowledge. And, thus, Act 590 does not have the advancement of empirical knowledge as a major effect. Here in brief is the argument for (23). Because creation science makes only religious and empirical claims, it makes no major contribution to other fields of study. Whence, Act 590 does not have the advancement of any other aim as a major effect.

Of course, even this line of argument is not without its problems. Perhaps it could be argued that exposing students to creation science would be useful in teaching them to criticize unwarranted empirical claims, and so maybe creation science could be made to serve some legitimate pedagogical purpose, though hardly the purpose envisaged by those who favored Act 590. Still, it seems more promising than the line of argument actually pursued in Judge Overton's Opinion. It is too bad that myopic focus on whether creation science is science led to the neglect of this and similar argumentative strategies in *McLean* v. *Arkansas,* and it is a shame that none of the expert witnesses brought such possibilities to Judge Overton's attention.

Conclusions

Scientists and their friends should derive little comfort from the outcome of *McLean* v. *Arkansas.* Victory was indeed achieved at the wholly unnecessary "expense of perpetuating and canonizing a false stereotype of what science is and how it works".[40] Philosophers of science may hope to derive from this a cautionary tale.

Like other academic specialists, philosphers of science run multiple risks when they bring their expertise into the policy-making arena. Three substantial ones spring to mind. One is the risk of failure to communicate. If the expert is utterly unable to communicate with the other participants in the policy-making process or to show how his or her expert opinion bears on the policy question under consideration, then he or she will fail to affect the outcome of the process. Clearly this was not a problem in *McLean* v. *Arkansas.* Judge Overton understood pretty thoroughly where Ruse's arguments were tending and how Ruse's conclusions could be applied to the question he had to decide. Another is the risk of being misunderstood. If the other participants in the policy-making process do not grasp all the nuances of the expert's opinion, they are likely to interpret it in a way that leads to mistaken inferences and to policies contrary to those the expert's opinion is intended to support or can reasonably be construed as supporting. Somewhat by accident, this turned out not to be the major problem in *McLean* v. *Arkansas* either. To be sure, Judge Overton did not appreciate the distinction between applying the criteria of testability and falsifiability at the level of individual statements and applying those criteria at the level of whole theories, and on that account he made some mistakes in arguing for the conclusion that creation science is not science. But Ruse arrived at the same conclusion by a different, though equally fallacious, route. And obviously Ruse's opinion that creation science is not science was intended to support and can reasonably be construed as supporting the decision to declare Act 590 unconstitutional. The third risk is that of misrepresentation. If the expert's views are not representative of a settled consensus of opinion in the relevant community of scholars, then policy based on those views will lack credibility within that community, and the members of that community are likely to regard such lack of credibility as discrediting the policy in question. This was the major problem in *McLean* v. *Arkansas.* Ruse's views do not represent a settled consensus of opinion among philosophers of science. Worse still, some of them are clearly false and some are based on obviously fallacious arguments. If they do not suffice to discredit the decision in *McLean* v. *Arkansas* among philosophers of science, it is only because it is at least arguable that Judge Overton made no mistake in reasoning to the conclusion that Act 590 fails to pass the first, and perhaps also the third, part of the three-pronged test. But, as I suggested at the end of the previous section, even this problem might have been avoided by a philosopher of science who focused his expert testimony on the real philosophical and scientific defects in creation science.

So I see no objection in principle to philosophers of science taking the applied turn by serving as expert witnesses. There are substantial risks, to be sure, and many philosophers of science would be reluctant to run them. The moral I draw from *McLean* v. *Arkansas* is that philosophers of science should not underestimate the risks but should proceed with care and caution to minimize them. One bad precedent, particularly one so extensively publicized and so apt to arouse passionate feelings, is already one too many.

Notes

1. Peter D. Asquith and Henry E. Kyburg, Jr., eds., *Current Research in Philosophy of Science* (East Lansing, Mich.: Philosophy of Science Association, 1979). In the present context, what is significant about this volume is not what it says, which is of high quality, but what is left unsaid.

2. William R. Overton, "Opinion in *McLean v. Arkansas*", *Science, Technology & Human Values* 7, no. 40 (Summer 1982): 29.

3. Larry Laudan, "Commentary: Science at the Bar—Causes for Concern", *Science, Technology & Human Values* 7, no. 41 (Fall 1982): 16.

4. Michael Ruse, "Response to the Commentary: *Pro Judice*", *Science, Technology & Human Values* 7, no. 41 (Fall 1982): 20.

5. Overton, "Opinion in *McLean v. Arkansas*", p. 40.

6. *Ibid.*, pp. 33–34.

7. *Ibid.*, p. 34.

8. *Ibid.*, p. 35.

9. See the translation of Genesis in *The New American Bible* (New York: P. J. Kenedy & Sons, 1970) and especially the explanatory footnote to Genesis 1: 2.

10. Overton, "Opinion in *McLean v. Arkansas*", p. 36.

11. *Ibid.*

12. *Ibid.*

13. *Ibid.*

14. *Ibid.*

15. *Ibid.*

16. *Ibid.*, pp. 36–37.

17. Laudan, "Commentary", p. 18.

18. Michael Ruse, "Creation Science is Not Science", *Science, Technology & Human Values* 7, no. 40 (Summer 1982): 73.

19. Stephen Jay Gould, *Hen's Teeth and Horse's Toes* (New York & London: W. W. Norton & Company, 1983) p. 256. The essay from which this sentence is quoted originally appeared in *Discover.*

20. *Ibid.*, pp. 384–385. The essay from which this passage is quoted originally appeared in *Natural History.*

21. *Ibid.*, p. 385.

22. Laudan, "Commentary", p. 17.

23. Ruse, "Response", p. 20.

24. See also Larry Laudan, "More on Creationism", *Science, Technology & Human Values* 8, no. 1 (Winter 1983): 37.

25. Ruse, "Response", p. 20.

26. Ruse, "Creation Science", p. 77.

27. *Ibid.*, p. 72.

28. *Ibid.*, p. 76.

29. *Ibid.*, p. 72.

30. *Ibid.*, p. 73.

31. *Ibid.*

32. Ruse, "Response", p. 21.

33. Ruse, "Creation Science", p. 73.

34. *Ibid.*

35. *Ibid.*, p. 75.

36. *Ibid.*

37. *Ibid.*, p. 74.

38. Nicholas Wade, "Madness in Their Method", *The New Republic* (June 27, 1983): 13–17.

39. Laudan, "Commentary", p. 18. For further discussion along the same lines, see Larry Laudan, "The Demise of the Demarcation Problem", *Physics, Philosophy and Psychoanalysis*, ed. by R. S. Cohen and L. Laudan (Dordrecht, Boston and Lancaster: D. Reidel, 1983), pp. 111–128.

40. Laudan, "Commentary", p. 19.

The Hermeneutic Construal of Psychoanalytic Theory and Therapy: An Ill-Conceived Paradigm for the Human Sciences

ADOLF GRÜNBAUM

1. Introduction

During the past fifteen years, the philosophers Paul Ricoeur (1970; 1974; 1981) and Jürgen Habermas (1970; 1971; 1973), as well as the psychoanalysts George Klein (1976) and Roy Schafer (1976), have put forward "hermeneutic" reconstructions of the Freudian corpus. These new construals are intended to supplant Freud's own perennial view of the psychoanalytic enterprise as a natural science[1] (S.E. 1933, 22:159; 1940, 23:158 and 282), a view also espoused by such latter-day influential analysts as Charles Brenner (1982:1–5).

Thus Ricoeur (1970:358) hailed the scientific liabilities of Freud's clinical theory, deeming the poverty of its scientific credentials a welcome ground for a "counter-attack" against those who deplore this very poverty. Indeed, quite recently Ricoeur (1981:259) endorsed anew Habermas's complaint that Freud had fallen prey to a portentous "scientistic self-misunderstanding". As Habermas would have it, this unfortunate error consisted in Freud's attribution of natural science status to his own clinical theory. According to Habermas (1971:246–252) and George Klein (1976:42–49), this misattribution was effected by *misextrapolation* from the would-be reduction of the clinical theory to a primordially scientific, neurobiologically inspired energy model of the mind.

In the spirit of the natural sciences, Freud maintained unswervingly that unconscious and conscious motives are each a species of the genus cause (S.E. 1909, 10:199; 1910, 11:38). Yet Klein (1976:56; also 12, 21) maintained that *bona fide* psychoanalytic explanations provide "reasons rather than causes" for human conduct. And, in the same vein, Roy Schafer (1976:204–205) avers that the adduced unconscious "reasons" multiply fail to qualify as causes.

In my book *The Foundations of Psychoanalysis* (1984), I gave a lengthy censorious appraisal of the most influential hermeneutic versions of psychoanalytic theory and therapy.

Here I shall challenge Paul Ricoeur's account in particular, which is avowedly inspired by his own philosophy of science no less than by his reading of Freud. To provide the wider setting for the issues, I now merely recapitulate some theses from my (1983) and (1984) that are directed against Habermas's and Klein's hermeneutic accounts of psychoanalysis, but without defending these particular theses here.[2]

1. Habermas and Klein contrived an exegetical legend by misdepicting the mature Freud's perennial *notion of scientificity* as ontologically reductive, such that the scientific status of the clinical theory is parasitic on that of Freud's energy model. Such a reading is mythic, since it runs afoul of explicit and definitive contrary texts (S.E. 1925, 20: 32–33; 1914, 14: 77). These writings show that during all but the first few years of his psychoanalytic career, Freud's touchstone of scientific status was avowedly methodological instead of ontologically reductive (Grünbaum 1983: sec. 1; 1984: Introduction, sec. 1).

2. Habermas's Hegelian "causality of fate" doctrine is, alas, merely a callow, causal blunder (Grünbaum 1983: 8–9). There is no such causality in the context of psychoanalytic etiology and therapy, any more than there is in physics. And ironically, this supposed causality boomerangs completely to boot (Grünbaum 1984: Introduction, sec. 2A).

Yet according to Habermas, the dynamics of psychoanalytic therapy exhibits Hegelian "causality of fate" rather than the causality of nature, since the patient's psychoanalytic "self-reflection" allegedly "dissolves", "overcomes", or "subdues" the very *causal connection* linking the pathogen etiologically to his neurosis (Habermas 1971: 256–257 and 271; 1970: 302 and 304).

3. Important laws in electromagnetic theory, and in other branches of physics, tellingly refute Habermas's contention that nomological explanations in the natural sciences are generically effected by recourse to "context-*free*" or *non*-historical laws of nature (Grünbaum 1983: 9–11; 1984: Introduction, sec. 2B).

By relying on this untutored claim, he asserted a fundamental contrast between law-based explanations in the natural sciences, and those furnished by the application of psychoanalytic generalizations to individual life histories in a personalized narrative (Habermas 1971: 272–273).

4. Habermas's thesis that the analyzed patient has privileged epistemic access to the validation or discreditation of psychoanalytic hypotheses completely begs the question (Grünbaum 1984: Introduction, sec. 2C). Moreover,

this attribution of cognitive primacy to the analysand is gainsaid by the cogent testability of Freudian assumptions without any recourse at all to the verdicts from patients in the treatment setting.

Yet Habermas avowed a salient contrast between the analytic patient's purportedly privileged epistemic access and the cognitive role of the observer in the natural sciences.

5. When Habermas published his (1971), he was oblivious to Pierre Duhem's insightful account of refutation in physics, given half a century earlier. Thus uninformed, Habermas relied on a detrimentally simplistic version of the invalidation of physical hypotheses to erect a pseudo-asymmetry between their discreditation and the logic of falsifying psychoanalytic interpretations.

But Freud's alleged commission of a "scientistic self-misunderstanding" (Habermas 1971: chap. 10) is claimed to be demonstrated by just the theses that I have shown to be untenable. Furthermore, this supposed self-misunderstanding was purportedly far-reaching, if only because it thwarted the recognition of psychoanalysis as a paradigmatically depth-hermeneutic mode of inquiry, as the only tangible example of a science incorporating methodical self-reflection, and as potentially prototypic for the other sciences of man.

Upon conjoining the conclusions above from my (1983) and (1984) with the scrutiny of Ricoeur's rendition of psychoanalysis below, the following lesson will emerge, I trust:

(i) Besides doing exegetical violence to the Freudian corpus, the most influential proposed hermeneutic reconstructions of that body of hypotheses rest squarely on untutored or anachronistic paradigms of the natural sciences.

(ii) *Qua* citadel of psychoanalytic apologetics, the hermeneutic reconstructions are a complete fiasco, to the discomfiture of those advocates who embraced them in the expectation of salvaging the psychoanalytic legacy by obviating the imperative to validate it scientifically.

(iii) The pretentious trappings of hermeneutic psychoanalysis (e.g., Ricoeur's semiotic figment "semantics of desire") cannot conceal that this supposedly humanistic implementation of the Freudian enterprise is just an investigative *cul-de-sac,* a negativistic ideological battle cry, showing no promise of becoming a fruitful prototype for the human sciences generally.

In view of the shoddiness of the arguments offered to hermeneuticize Freud, I find it difficult to avert a pyschological conjecture as to the ideological inspiration of their proponents: They seek to *legitimate* (a) the methodological *absolution* of the study of human ideation from the evidential stringency governing the validation of theories in the standard empirical sciences and (b) an ontological demarcation between mentation and other natural processes, so as to vindicate the desired foregoing probative dispensation. To my mind, this motivation is likely to be political, religious, or the understandable desire to safeguard a lifetime professional investment in the practice of psychoanalytic treatment. For in the face of my scrutiny of the hermeneutic construal, I cannot see what other stakes its champions could plausibly have in its promulgation or endorsement.

2. Ricoeur's Encapsulation of the Purview of Freudian Theory within the Clinical Milieu

Ricoeur sets the stage for his proposed hermeneutic reconstruction by flatly truncating the domain of occurrences to which psychoanalytic theory is to be deemed relevant. Thus, he immures its substantive purview within the *verbal* productions of the clinical transaction between the analyst and the patient. Its subject matter, we are told, is "analytic experience [in that dyadic setting], insofar as the latter operates in the field of speech" (1970: 375). In this way, he stipulates at the outset that "the ultimate truth claim [of psychoanalytic theory] resides in the case histories" such that "all truth claims of psychoanalysis are ultimately summed up in the narrative structure of psychoanalytic facts" (1981: 268; see also 38 and 248).

Once the domain of relevance of Freud's theory is held to be coextensive with "a work of speech with the patient" (Ricoeur 1970: 369), even the analysand's *non*-verbal productions are excluded from its scope. But, as I shall argue, Ricoeur's circumscription (1981: 248) of the domain of "facts"—or objects of knowledge—that the pyschoanalytic corpus is declared to codify is a wanton mutilation of its range of relevance. In fact, his identification of the clinical utterances as the very object of psychoanalytic inquiry will now turn out to be unavailing to his aim of absolving psychoanalysis from the imperative to seek scientific validation.

Ricoeur blithely ignores at the outset that presumed observations in physics are already theory-laden and are then further interpreted theoretically. Thereby he then miscontrasts the psychoanalyst epistemically with the natural scientist in the garb of the *behaviorist* psychologist: "Strictly speaking, there are no 'facts' in psychoanalysis, for the analyst does not observe, he interprets" (Ricoeur 1970: 365). In truth, well before the appearance of Ricoeur's (1970), Popper, W. Sellars, and N. R. Hanson—not to mention Kant—had already discredited the crude observation-theory dichotomy, which Ricoeur (1981: 247–248) uncritically assumes for the natural sciences. Thus, to the serious detriment of his hermeneutic case, he overlooks that insofar as the interpretive activity of an observer militates against the existence of "*data*" — i.e., of "facts" to be explained — the received natural sciences are on the same epistemic footing as the interpretative clinical inferences drawn by analysts.

Having run afoul of this commonplace, he also invokes his peremptory confinement of the purview of psychoanalytic theory to the patient's speech from the couch. Thereupon, he draws the following conclusions: (i) Unlike the behavioral "facts" of scientific psychology, "facts in psychoanalysis are in no way facts of observable behaviour" (1981:248), and (ii) the epistemic scrutiny familiar from the empirical sciences cannot intrude upon the hermeneutic construal of Freudian theory.

As partial grounds for (i), he depicts Skinnerian behaviorism as ontologically and epistemically prototypic for *scientific* psychology. Yet his use of just that paradigm is illicit and tendentious, if only because *cognitive* scientific psychology, for example, countenances *intra*-psychic states, no less than psychoanalysis does (Maxwell & Maxwell 1972). Thus it emerges that Ricoeur's claims (i) and (ii) are ill-founded, even if one were not to cavil at his ontological shrinkage of the domain of psychoanalytic theory into the straitjacket of the dyadic "work of speech with the patient" in the treatment setting. But that encapsulation is itself an unwarranted ideological maneuver, as is about to become apparent.

True enough, psychoanalysts generally regard their many observations of the patient's verbal and non-verbal interactions with them in the treatment sessions as the source of findings that are simply peerless *as evidence*, not only heuristically but also probatively (Jones 1959, vol. 1:3). But this avowed *epistemic* tribute to the evidence garnered in clinical investigations is a far cry from limiting the very purview or subject-matter of the entire theory to the phenomena of the clinical transaction, let alone to "the work of speech with the patient". Nor is such stunting justified by the platitude that Freud intended the clinical findings to be *included within* the explanatory scope of his system of hypotheses. For his etiologic hypotheses purportedly explained generically why people at large acquire neuroses, *regardless of whether they are ever treated psychoanalytically or not.* Thus Freud claimed to have given an etiologic account of why some people become paranoiacs, even if — like the judge Schreber (S.E. 1911, 12: 3–82) — they are never seen by an analyst.

By the same token, he claimed to have illuminated why, even among the *unanalyzed*, the personality traits of obstinacy, orderliness, and parsimony tended to cluster together and deserved the etiologic label of "anal character". Moreover, as against Ricoeur, no less than against Habermas, it is germane that the psychoanalytic etiology of paranoia would explain an epidemiologic decrease in the incidence of paranoia by pointing to the decline of the taboo on homosexuality (Grünbaum 1983: 12–13). And this explanation would or could be given quite apart from any concern for data from the analytic treatment setting. Indeed, just as the intersubjective epidemiologic testability of the psychoanalytic etiology of paranoia gainsays Habermas's epistemology, so also it contravenes Ricoeur's dichotomy of validation between psychoanalytic theory and academic scientific psychology. Yet Ricoeur (1974: 186) insists that "psychoanalysis does not satisfy the standards of the sciences of observation, and the 'facts' it deals with are not verifiable by multiple, independent observers. . . . there are no 'facts' nor any observation of 'facts' in psychoanalysis but rather the interpretation of a narrated history."

The therapeutic dynamics depicted in psychoanalytic theory is hardly restricted to speech acts on the analyst's couch or in his (her) office. Yet there is an obvious sense in which Freud's therapy may be and has been dubbed a "talking cure". As he has told us (S.E. 1910, 11: 13 and 21), Breuer's first patient, Anna O., coined just this label for the proto-psychoanalytic treatment she had received from Freud's mentor. But it is dull to be told by Ricoeur that "analysis qua 'talking cure'" is a "closed field of speech" (1970: 369). And thus, it hardly vindicates his willful impoverishment of the domain of "facts" addressed by Freud to adduce that "psychoanalysis [qua 'talking cure'] is itself a work of speech with the patient . . . it is in a field of speech that the patient's 'story' is told." Most recently, Ricoeur issued the mild disclaimer that "there is no need to insist here on the *talk-cure* character of psychoanalysis" (1981: 248). But right after this slight demurrer, he hastened to maintain that the "screening [of thoughts and feelings] through speech in the analytic situation also functions as a criterion for what will be held to be the object [of knowledge] of this science". And the verdict of that criterion is purportedly the identification of "only that part of [analytic treatment] experience which is capable of *being said*" (1981: 248) as the object of psychoanalytic knowledge.

The psychoanalytic dream theory springs the confines of Ricoeur's speech-acts domain no less than Freud's etiologies do. As Ricoeur would have it, "it is not the dream as dreamed that can be interpreted, but rather the text of the dream account" (1970: 5). And without telling us just what is to be understood by the term "know", he avers: "We know dreams only as told upon awakening" (1981: 248). Does "know" here require awareness of the manifest content *and* of its being fancied? If so, why must people *verbalize* the memories of their dreams, if they are to "know" them in waking life?

But even if they must do so, how would this show that the domain of psychoanalytic dream theory is devoid of dreams as actually dreamt and is confined to verbalized memories of them? Why does the domain of physical theory comprise elementary particles rather than *only*, say, the tracks they leave in Wilson cloud chambers or other registration devices, whereas the domain of the dream theory is to contain *only* verbalized memories of dreams? What, besides an imported ideological objective, prompted Ricoeur to shrink the *subject-matter* of Freud's wish-fulfillment theory, which offers repressed infantile motives for dreams as dreamt during sleep, into mere verbal dream-reports during waking life?

Perhaps Ricoeur's contrived ontological stultification of the psychoanalytic dream theory was abetted by the unsound inference that the dreamer's verbalized recollections of his (her) dreams are the domain of relevance for any theory of dreams, simply because these utterances are presumed to be the *epistemic point of departure* for such a theory. Of course, Freud himself eschewed this inference, although he countenanced its premise. In his view, the epistemic *terminus a quo* of dream research was indeed an assortment of manifest dream contents, as fallibly recalled and reported by the dreamer, perhaps even after some editing (S.E. 1916, 15: 84–85). Yet while acknowledging these defects in the "data", he put them into perspective: They did not militate against the generic explanation of dreams as presumed to have been dreamt by the unanalyzed vast majority of mankind and by the analyzed alike.

It is salutary in this context to appreciate that what astronomers endeavor to explain, in the first instance, is not the visual impressions they have of the celestial occurrences, but the celestial events *retrodicted* from these impressions, or the events whose earlier occurrence their impressions are taken to betoken. And just as dream reporting may well be inaccurate, so also the accuracy of astronomical observations has been lessened by the earth's atmosphere, for example. Small wonder, therefore, that what Freudian theory tries to explain, in the first instance, is not the verbal dream-*reports* made by dreamers to analysts or others, but rather the manifest dream-contents inferred from the dreamers' subsequent mnemic impressions. Nay, Freud claimed to have explained why people dream at all, in addition to purporting to illuminate generically why they dream what they dream, whether they are psychoanalyzed or not.

There are further psychoanalytic hypotheses that tell against Ricoeur's truncation of their purview. Thus, according to Freud, the so-called "transference" phenomena become important in the treatment-transaction, precisely because the encounter between the analyst and the patient poignantly *instanti-*

ates processes that are operative in all of us, even if we never enter an analysis in our lives. Thus, the unanalyzed, no less than those who embark on a psychoanalysis, are held to *transfer* to their adult interpersonal relations unconscious conflict-laden attitudes (e.g., Oedipal feelings) originally entertained in childhood toward important figures (e.g., a parent). And by being infantile, the unwittingly transposed dispositions are inappropriate to adult life situations. Hence by carrying them over into his (her) interpersonal relations, the mature person reacts to others as if they were figures from his (her) distant early past, thereby often misattributing alien motives and traits to them. In short, in all of mankind, the "transference phenomena" purportedly distort interpersonal relations. And it is on the strength of being deemed universally operative that the hypothesized transference is claimed to explain the interaction of psychoanalytic patients with their doctors. Hence the hypothesized transference furnishes yet another case against Ricoeur's contrived and tendentious attempt to shrink Freud's theory into the mold of analytic experience on the couch.

Moreover, this ideological surgery on the psychoanalytic corpus hardly coheres with Ricoeur's belated recognition of the actual scope of *causal explanation* in Freudian theory. He rightly deems causal claims to be intrinsic to this corpus of hypotheses. Yet, as we shall now see, the range of relevance of just these claims is far wider than that of "the hermeneutics of self-understanding" (1981:264) purportedly employed within the treatment setting.

On the one hand, Ricoeur told us most recently that "facts in psychoanalysis are in no way facts of observable behaviour" (1981:248). But, in the very same chapter, he went on to gainsay this dictum:

> What is remarkable about psychoanalytic explanation is that it brings into view motives which are causes. . . . In many ways his [Freud's] explanation refers to "causally relevant" factors. . . . All that is important to him is to explain . . . what in behaviour are "the incongruities" in relation to the expected course of a human agent's action. . . . It is the attempt to reduce these "incongruities" that . . . calls for an *explanation* by means of causes. . . . To say, for example, that a feeling is unconscious . . . is to say that it is to be inserted as a causally relevant factor in order to explain the incongruities of an act of behaviour. . . . From this . . . it follows . . . that the hermeneutics of self-understanding take the detour of causal explanation. (pp. 262–264)

Here Ricoeur evidently recognizes that psychoanalytic explanations are both *causal* and are intended to illuminate various sorts of behavior. If so, then their validation, if any, will have to be of a kind *appropriate* to these avowed features. But how does Ricoeur envision the appraisal of the psychoanalytic codification of a purported causal connection? As I have argued in detail in Part II of my (1984), the demonstration of the *causal* relevance of various sorts of repressions cannot be effected *within* the treatment-setting by such clinical methods as free association, which are endemic to the Freudian enterprise. Instead, the establishment of a causal connection in psychoanalysis, no less than in "academic psychology" or medicine, has to include reliance on received modes of empirical inquiry that were refined from time-honored canons of causal inference pioneered by Francis Bacon and John Stuart Mill.

Now, the imperative to furnish cogent evidence for the purported causal linkages invoked to explain the patient's case history is not lessened by the injunction (Ricoeur 1981: 266–268) to fulfill the "narrativity criterion" as well. The latter requires that "the partial explanatory segments of this or that fragment of behaviour are integrated in a narrative structure" reflecting the individual analysand's etiologic life history (p. 267). But, as Ricoeur emphasizes, the psychoanalytically reconstructed scenario not only must be a *"coherent story"* (p. 267) — made "intelligible" by the explanatory [etiologic] segments — but must also aspire to being true, rather than merely persuasive and therapeutic (p. 268). Quite properly, therefore, he enjoins that "we must not give up our efforts to link a truth claim with the narrativity criterion, even if this claim is validated on a basis other than narrativity itself" (p. 268).

Indeed, he elaborates (pp. 268–269) on "what makes a narration an explanation in the psychoanalytic sense of the term" as follows: "It is the possibility of inserting several stages of causal explanation into the process of self-understanding in narrative terms. And it is this explanatory detour that entails recourse to non-narrative means of proof." Significantly, he adds that the three levels over which these means of proof are spread include "the level of law-like propositions applied [*mirabile dictu!*] to typical segments of behaviour (symptoms, for example)" (p. 269). Yet earlier in the same chapter, Ricoeur had adduced "even [neurotic] symptoms" in support of his claim that "facts in psychoanalysis are in no way facts of observable behaviour" (p. 248). For there, he had declared that "even symptoms, although they are partially observable, enter into the field of analysis only in relation to other factors verbalized in the 'report'". The conscientious reader will be forgiven, I trust, for wondering whether Ricoeur himself has decided just what he wants to maintain.

At any rate, it emerges that the causal hypotheses on which psychoanalytic explanations are predicated must be appraised, after all, by none other than the methods of "academic psychology" and/or of the natural sciences! But Ricoeur had argued — in Section I of the same chapter — that just these modes of assessment ought not to be countenanced. As he told us, they are inappropriate, because they differ *toto genere*, if not *toto caelo*, from the standards governing "the hermeneutics of self-understanding".

Indeed, in Ricoeur's commentary on the editorial introduction by J. B. Thompson, the translator of his 1981 book, he (pp. 32–40, esp. p. 38) apparently endorses the methodological dichotomy enunciated there on his behalf by Thompson. Significantly, the latter declared (1981: 7):

> In response to those critics who contend that Freud's theory does not satisfy the most elementary criteria of scientificity . . . Ricoeur maintains that all such contentions . . . betray the very essence of psychoanalysis. For the latter is not an observational science dealing with the facts of behaviour; rather it is an interpretative discipline concerned with relations of meaning between representative symbols and primordial instincts. Thus, psychoanalytic concepts should be judged, not according to the exigencies of an empirical science, but "according to their status as conditions of the possibility of analytic [treatment] experience, insofar as the latter operates in the field of speech."

And when speaking of "the question of proof in Freud's psychoanalytic writings", Thompson points out further (Ricoeur 1981: 24) that "Ricoeur's current [1981] approach to this question reveals a shift away from his earlier [1970] work" *Freud and Philosophy:* "His starting point now [in 1981] is the analytic situation, which determines what counts as a 'fact' in psychoanalysis."

Ricoeur's depiction of the epistemology of psychoanalysis is thus fundamentally flawed. No wonder, therefore, that his account altogether fails to heed the methodological import of the causal nature of psychoanalytic explanation. As he acknowledges (1981: 263), Freud gives explicitly causal explanations of "the origin of a neurosis". But once this is granted, Freud's modification of his erstwhile seduction etiology of hysteria can now be shown to be quite unavailing to Ricoeur's indictment of "scientistic" psychoanalysis.

3. Are Natural Science Modes of Explanation Inapplicable to Pathogenic *Phantasies*?

As heralded in an important 1897 letter to Wilhelm Fliess, Freud (1954, Letter #69, pp. 215–218) was driven to invoke *fancied* childhood seductions after first having deemed actual ones to be the pathogens of hysteria. But even when he still thought *bona fide* seductions to have been the etiologic factor, their hypothesized pathogenicity was, of course, held to be crucially mediated causally by the *psychic* trauma of the child's *experience* of these sexual episodes. This etiologic mediation of exogenously actuated, yet *intra-psychic* processes lends perspective to a Freudian passage adduced by Ricoeur (1981: 250) in the mistaken belief that it furnishes ammunition to his anti-"scientistic" cause. In the pertinent passage, Freud pays etiologic tribute to childhood seduction phantasies as follows:

> It remains a fact that the patient has created these phantasies for himself, and this fact is of scarcely less importance for his neurosis than if he had really experienced what the phantasies contain. The phantasies possess *psychical* as contrasted with *material* reality, and we gradually learn to understand that *in the world of the neuroses it is psychical reality which is the decisive kind.* (S.E. 1917, 16: 369)

But Ricoeur can extract no philosophic capital from the hypothesis that the purportedly etiologic process is now held to be actuated causally by autochthonous psychic events rather than exogenously and reactively, in response to events having "*material* reality". For the fancied character of the seductions hardly obviates the imperative to validate the pathogenicity attributed to these phantasies. Yet Ricoeur overlooks in this context—just as Roy Schafer (1976: 204–205) ignores in another connection—that no matter whether the seductions are actual or only fancied, the exponent of their etiologic role cannot be absolved from giving cogent evidence for that causal attribution. And—as Part II of my (1984) shows—the procurement of such support, in turn, requires precisely the methods decried by the hermeneuticians.

In a much publicized book, J. M. Masson (1984) overlooks the same point as Ricoeur does, though Masson rejects fancied seductions as the pathogens, and instead credulously assigns that etiologic role to *actual* episodes of sexual child abuse. Hence Masson sees Freud's repudiation of his erstwhile seduction etiology as a fateful error, which grievously misdirected all of his subsequent theorizing. And he contends that a failure of nerve, rather than contrary evidence, prompted Freud's disavowal. But, whatever the merits of that accusation, Masson's rehabilitation of actual seductions as the pathogens is etiologically unfounded, as is the pathogenicity of fancied ones.

Thus it is a corollary of chapter 3 of my (1984) that just as the method of free association is incompetent to warrant the pathogenicity of truly occurring childhood seductions, so also this method cannot attest that imagined ones were etiologic. Indeed, as chapter 8 (1984) makes clear, the relevant effect of Freud's replacement of actual childhood seductions by merely fancied ones as pathogens is only to make the task of validating their etiologic relevance much harder! Yet Ricoeur (1981: 251) lulls himself into the ill-founded belief that by contrast to academic psychology, "Psychoanalysis deals with psychical reality and not with material reality. So [!] the criterion for this reality is no longer that it is observable [however indirectly, as in physics], but that it presents a coherence and a resistance comparable to that of material reality" (1981: 251). But where in all this is there even a hint as to why we should believe that children who assumedly concoct seduction figments then become prone to developing hysteria *because* they entertain such phantasies? And furthermore, may it not be that the provenance of the seduction phantasies themselves is not entirely intra-psychic but involves "material reality"?

On the other hand, if there is actually an etiologic connection between such phantasies and hysteria, then the *truth* of its affirmation is obviously not impugned in the least by the *fictitiousness* of the fabricated seductions. Ricoeur transmogrifies this patent fact into "the major difficulty facing the truth claims of psychoanalysis" (1981: 266). Let us even grant him that "what is psychoanalytically [etiologically] relevant [in this context] is what a subject makes of his fantasies", while being duly wary of the misleading locution that the patient "makes" hysteria pathogenically out of his seduction phantasies. And let us countenance, furthermore, "giving wider rein to the liberation of fantasising, to emotional development, and to enjoyment than Freud wanted to do". How then, I ask, is it "undoubtedly" the case that we are thereby "breaking the [unspecified] bond between veracity [truth-telling] and truth" (p. 266)? Not insisting on veracity does not affect whatever *bond* there is between any existing veracity and truth, does it?

When a psychiatrist allows a psychotically paranoid patient much freedom to express his delusions — or even panders to them — is the doctor thereby jeopardizing "the bond between veracity and truth"? Or is he merely not insisting on veracity? Unencumbered by such prosaic considerations, Ricoeur tries the reader's patience further by allowing: "Nevertheless, I think that there is still something to be sought in the truth claim made from the perspective of the proper use of fantasies" (p. 266). Thereupon he launches into a disquisition as to why this should be so.

Ricoeur himself seems to have conceded, albeit only implicitly and unwittingly, that etiologic hypotheses require validation *qua* being causal, regardless of whether the actuating pathogen is held to be an autochthonous psychic event — e.g., a seduction *phantasy* — or a response to events having "material reality", such as an actual seduction experience. As he acknowledges, any reconstruction of psychoanalysis must include its "task of integrating an explanatory stage", which "keeps psychoanalysis from constituting itself as a province of the exegetical disciplines applied to texts — as a hermeneutics" (1981: 261).

And this explanatory imperative, in turn, "requires that psychoanalysis include in the process of self-understanding operations that were originally reserved for the natural sciences" (p. 261).

Indeed, the "non-narrative" generalizations employed in psychoanalytic explanations are avowedly

> already present in the ordinary explanations of individual behaviour; alleged motives — for example, hate or jealousy — are not particular events, but classes of inclinations [types of dispositions] under which a particular action is placed in order to make it intelligible. To say that someone acted out of jealousy is to invoke in the case of his particular action a feature which is grasped from the outset as repeatable and common to an indeterminate variety of individuals. . . . So to explain is to characterize a given action by ascribing to it as its cause a motive which exemplifies a class. (Ricoeur 1981:269)

But, once this is recognized, as indeed it should be, its import completely undermines the hermeneutic denunciation of natural science modes of explanation in psychoanalysis. For it then becomes quite unavailing to adduce, in the words of George Klein (1976:26), that "the central objective of psycho-analytic clinical explanation is the *reading of intentionality;* behavior, experience, testimony are studied for meaning in this sense". Instead, the hermeneutic denunciation of natural science modes of causal explanation in psychoanalysis is undercut by the *import* of the following Freudian thesis: There are, in all of us, repressed and even conscious motives that are not only first discerned to exist by analysts, but are also the *causes* of the sorts of behaviors, thoughts, feelings, etc. purportedly explained by them. *In the context of the psychoanalytic clinical theory of psychopathology, the ontological identity of unconscious ideation* qua *being mental rather than physical hardly robs it of its hypothesized causal role.* For, as I have documented elsewhere (Grünbaum 1983: sec. 1), this Freudian theory of repression emphatically abjured an exclusively physicalistic construal of the attribute of causal relevance. Thus, within the psychic world depicted by Freud, it is simply wrongheaded to maintain that, in virtue of being mental, repressed ideation cannot function as a motivational *cause.* As well aver the inanity that because an explosion involves natural gas, this eruption cannot qualify as a cause of the collapse of a building and of the deaths of its occupants.

Yet such advocates of hermeneutic psychoanalysis as George Klein (1976: 43) and Roy Schafer (1976:204–205) have succumbed to an equally mistaken belief. In their misconstrual of Freud's vision, such *causes* as there are of human conduct cannot be genuinely motivational in the mental fashion of unconscious or conscious motives. And their incongruous notion has fostered the pernicious myth that precisely insofar as explanations in psychoanalysis are indeed motivational or supply unconscious "reasons" for our actions, they cannot be a particular *species* of causal explanations (Grünbaum 1984: Introduction, sec. 4). But, as Ricoeur came to appreciate to his great credit (1981: 262–263, 269), those who draw this conclusion are in effect *repudiating* Freud's clinical explanations of neuroses, dreams, and parapraxes, although they see themselves as giving a philosophic explication of them! For in his clinical thery of repression, unconscious motives are deemed to be explanatory precisely because they are held to *engender* these various manifestations of their existence by being causally relevant to them. And the issue here is not whether it is sound to explain human actions causally, but only whether psychoanalysis can be reconstructed short of emasculation without doing so.

Thus, the psychoanalytic quest for the so-called "meaning" of neurotic symptoms, dreams, and slips turns out to be predicated on a crucial two-fold presupposition. First, the *concealed* "intentionality" postulated by Freud's theory does in fact exist covertly, thereby first turning these various facets of human life into targets for being psychoanalytically deciphered. After all, without the existential assumption that *there are* unconscious processes holding the key to the significance to be fathomed, it is simply illusory to investigate psychoanalytically the arcane "meaning" of the life events in question. Second— as against the stated misconceptions—the putative repressed ideation derives its *explanatoriness* in psychoanalysis from its hypothesized *causal* role, an elucidative pedigree "originally reserved for the natural sciences" (Ricoeur 1981: 261). But just this attribution of various sorts of causal efficacy to unconscious mentation cries out for cogent vindication. Hence the very quest for the veiled "meaning", which psychoanalytic explanation is expected to disclose, *cannot redeem its avowed promise* without prior reliance on *methods of inquiry and validation* that hermeneuticians proclaim to be quite inappropriate: exactly those cognitive procedures that they are wont to categorize as endemic to the natural sciences. By the same token, insofar as these partisans do countenance causal explanations in psychoanalysis, their whole hermeneutic enterprise is tacitly parasitic on an epistemology they inveterately profess to decry.

Ricoeur is no less guilty of this legerdemain than others. For, as we saw, he emphatically denounces the logic of validation familiar from the natural sciences as alien to Freud's clinical theory. And he does so by misassimilating these sciences untutoredly to his mythic notion of an "observational science" (1981, chap. 10, sec. I; 1970: 358–375). Regrettably, this baneful error is not mitigated by his admission (1981: 261) that the causal explanations furnished by psychoanalysis gainsay the confinement of the Freudian edifice to a mere "province" of *exegetic* (textual) hermeneutics.

Let us now use a poignant illustration to put into bolder relief the extent to which the entire hermeneutic enterprise is *ill-conceived* in psychoanalysis, unless it provides adequate scope for validating the causal imputations implicit in the hidden "meaning" it purports to fathom.

Suppose that the thought-fragmentation and cackling exhibited by a certain class of schizophrenic women were taken to betoken the witchcraft cunningly practiced by them and/or their unawareness of being satanically possessed. Those claiming to have acceptable grounds for such an attribution would, of course, consider themselves justified when summoning a shaman or an exorcist. By supposedly conjuring up the possessing spirits or the like, this therapist would then fathom hermeneutically the conjectured "hidden meaning" of the babble from these unfortunate women. But the rest of us will be forgiven for telling the shamans and exorcists that we deem their hermeneutic quest and ensuing revelations to be ill-conceived. For they have conspicuously failed to validate cogently the bizarre causal imputations on which their "clinical investigations" of "meaning" and their therapy are predicated.

True, their incantations may even favorably affect the schizophrenic symptoms, if only temporarily. But even if these therapeutic results were impressive, we would hardly credit the shamans or exorcists with having unraveled the otherwise elusive "meaning" of the witchcraft or possession symptoms. For we reject their underlying causal ontology, which determines what kind of "meaning" their quest will "uncover". And thus we dismiss their "reading of intentionality" as altogether chimerical.

Likewise, if a new school of self-styled archaeologists were to announce an array of retrodictive inferences, based on exceedingly dubious causal hypotheses, we would likewise withhold our assent from their asseveration of previously elusive "archaeological meaning".

Turning to psychoanalysis, the noted analyst B. B. Rubinstein has expressed the same animadversion for the hermeneuticians in the context of dream theory. In a section "Explanation in Terms of Meaning and Causal Explanation", Rubinstein (1975: 104–105) wrote incisively:

> When we interpret the meaning of a dream symbol we presuppose the actual occurrence of the very processes we posit to explain causally the production of the symbol. Clearly . . . no matter how apt an interpretation of a symbol in terms of its [purported latent] meaning, if the processes by which symbol formation is explained are improbable, we have no alternative but to discard the interpretation. . . . to an interpreter [e.g., the analyst] the *meaning* of a

dream symbol is what the symbol, unbeknownst to the dreamer, is taken to *signify* . . . however, either the thus signified is also what has given rise to, i.e., *caused* (or contributed to cause) the occurrence of, the symbol, or the 'symbol' cannot be said to signify it. We will note that if a symbol is justifiably said to signify something then it can also be said to *represent* this something.

One salutary corollary of this cogent statement is the following: no matter how strong the *thematic affinity* between a conjectured repressed thought and a maladaptive, neurotic action, this "meaning kinship" does not itself suffice to attest that the hypothesized repression is "the hidden intentionality" behind the given behavior. For thematic affinity alone does not vouch for etiologic lineage in the absence of further evidence that a thematically kindred repression actually *engendered* the behavior! And if there is no warrant for inferring the operation of the repression from the patient's conduct, it is only misleading to assert that the former is "the meaning" of the latter or that the latter "signifies" the former. For the behavior in question does not in fact bespeak the existence of the repression, and hence does not justify reading the latter into it. But by using the weasel word "meaning" to render what—so far as the evidence goes—is *only* a thematic affinity, the putative hermeneutician illicitly creates the semblance of having uncovered a "hidden intention" after all.

Even analysts who are not avowed hermeneuticians are prone to what might be dubbed "the thematic affinity fallacy". The diagnosis made in a case discussed in some detail at the end of chapter 4 of my (1984) will now serve to illustrate a cognate pitfall. As documented there, an analyst had a woman patient who had been unduly concerned about her general appearance, notably her skin. There is, of course, some thematic affinity between the dermatological deficit resented by the patient and the phallic deficit which, according to Freud's hypothesis of female penis envy, is resented by little girls who are held to envy their brothers the male anatomical endowment. Besides, there is also resentment in both cases. But assuming that there are grounds for attributing penis envy to little girls, this alone is a far cry from showing that their sense of anatomical "defect" is *also pathogenic*. Yet, in the case discussed late in chapter 4 of my (1984), the analyst diagnosed penis envy to be the hidden intentionality behind the female patient's discontent with her skin and general appearance, after having orchestrated her assent to the imputation of such envy.

Interestingly, there are also erstwhile behaviorists who overlooked that conversion to hermeneutic psychology does not provide absolution from the stated imperative to validate the causal hypotheses tacitly invoked by their avowed enterprise. Insouciantly endorsing psychoanalytic imputations of hidden motives, A. Gauld and J. Shotter (1977) say nothing about the grounds a hermeneutic psychologist would have — beyond thematic affinity — for claiming, for example, that unconscious aggression toward the father was the motivational cause of a patient's particular behavior. Thus, in their view, it is one mission of the hermeneutic psychologist "to elucidate by any method that he can the 'meanings' which the actions had for the agents. This task is completed when after psychotherapy the patient comes to understand that his strange action was a symbolic piece of aggression against his father" (p. 80). But what are their grounds for supposing that any of the methods countenanced by their hermeneutic doctrine have any promise to uncover "the [unconscious] 'meanings' which the actions had for the agents"?

As I pointed out in my (1983: 20–21), there is good evidence from cognitive psychology that, even in the case of *consciously* motivated behavior, the agent does not enjoy privileged epistemic access to the discernment of the motives for his (her) various commonplace actions. Indeed all too frequently, the purportedly introspective ascertainment of conscious motives is demonstrably wrong. Thus, when an agent offers a motivational explanation for his own actions, he does so — just like outside observers — by invoking *theory-based* schemata endorsed by the belief-system to which he adheres. *A fortiori*, in the case of *unconsciously* engendered behavior, the patient and the analyst alike are plainly drawing on psychoanalytic theory when purporting to read the previously unrecognized "meaning" of the patient's "strange action".

How then can either of them be held to know — on Gauld and Shotter's showing, short of resorting to the forsaken extra-hermeneutic methods of validation — that this "meaning" was unconscious aggression toward the father at all? For the purpose of effecting a change in the agent's odd behavior, it may perhaps be quite irrelevant whether the therapist's reading of the instigating unconscious motive is mythic. But Gauld and Shotter use the phrase "the [unconscious] 'meaning' which the action had for the agent". And their use of the past tense suggests their existential assumption of a motive that was actually operative even before the patient entered therapy, rather than of a putative motive that need only become retrospectively believable for him under the influence of treatment. Yet if so, then it is anything but immaterial to their declared elucidatory objective that the analyst's motivational imputations have better credentials than mere fancy. Nonetheless, they beg the question by side-stepping this vital issue. For they conclude genially: "Once the formerly 'obscure' meanings of these various actions have been made clear, so that the actions are now as well understood as the commonplace actions which everyone can comprehend, this stage of the hermeneutical psychologist's task is over" (p. 81).

Other hermeneuticians have sought to immunize their undertaking against the risks of causal imputations by reconceptualizing, more radically than Ricoeur or Gauld and Shotter, the very aims of the psychoanalytic elucidation of "meaning". This stratagem has likewise had much appeal to a good many analysts. For, faced with the bleak import of sceptical indictments of their legacy, they are intent on salvaging it in some form. Hence some of them will be understandably receptive to a rationale that promises them absolution from their failure to validate the cardinal hypotheses of their clinical theory, a failure I demonstrate in depth in Part II of my (1984). Be of stout heart, they are told, and take the radical *hermeneutic* turn. Freud, they learn, brought the incubus of validation on himself by his scientistic pretensions. Abjure his program of causal explanation, the more drastic hermeneuticians beckon them, and you will no longer be saddled with the harassing demand to justify Freud's causal hypotheses. One such hermeneutic advocate illustrated this repudiation of causation as follows: "The meaning of a dream does not reside in some prior latent dream [content, as Freud had claimed], but in the manifest dream and the analysand's associations to it" (Steele 1979: 400). Michael Moore (1983: 49) has lucidly explicated this posture before vigorously rejecting it. As he puts it, the claim is

> that Freud was not really explaining how the particular dream occurred . . . he was . . . not discovering its (motivational or nonmotivational) causes. On this account, the rationalizing but noncausal wishes discovered after the dream by free association have nothing to do with producing the dream; they are after-the-fact discoveries made by juxtaposing the manifest content of the dreams with the material produced by free association — an interpretive technique that may tell you something about yourself, but nothing at all of what caused your dreams.

More generally, the blandishments of this renunciatory stance include the comforting assurance that the practicing analyst is immune to the taunts of critics who dispute the cost-effectiveness of his therapy. And even for the Freudian psychohistorians, the hermeneutician has the glad tidings that henceforth they can hold their heads high as protagonists of a newly legitimated kind of humanistic discipline. In short, the claim is that the challenge to provide validation of the causal propositions has been obviated, and that the continuing demand for it has therefore become an anachronism, cherished only by positivist fanatics.

As for this sort of hermeneutics *without* causation, I have assessed it in my (1983: sec. 4, pp. 24–26): "It is a nihilistic, if not frivolous, trivialization of Freud's entire clinical theory. Far from serving as a new citadel for psychoanalytic apologetics, the embrace of such hermeneuticians is, I submit, the kiss of death for the legacy that was to be saved."

4. Does Freud's Theory of Repression Furnish a "Semantics of Desire"?

In psychoanalytic theory, both full-fledged neurotic symptoms and mini-neurotic ones (e.g., manifest dream contents, Freudian slips, jokes) are seen as *compromise*-formations, products of the defensive conflict between the repressed ideas and the repressing ones (S.E. 1896, 3: 170; 1917, 16: 358–359). *Qua* compromise-formations, *symptoms* have traditionally also been viewed as "symbols" of the repressed, but in the altogether *non*-semantic sense of being *substitutive* formations affording *replacement* satisfactions or outlets. Yet it would be an error to suppose that, at least more often than not, there is *thematic affinity* between the ideational or affective content of a repression, and the symptoms that "symbolize" it substitutively by providing an outlet for that particular content. To be sure, Breuer's first patient, Anna O., who was severely averse to drinking water, had presumably become thus hydrophobic by strangulating and repressing the disgust she had experienced earlier at the sight of seeing a dog lapping water from a companion's drinking glass (S.E. 1895, 2: 34). But, as Breuer and Freud emphasized at the outset (S.E. 1893, 2: 5), "the typical hysterical symptoms" do not have any apparent thematic connection at all to their presumed pathogens:

> It consists only in what might be called a "symbolic" relation between the precipitating cause and the pathological phenomenon — a relation such as healthy people form in dreams. For instance, a neuralgia may follow upon mental pain or vomiting upon a feeling of moral disgust. . . . In still other cases it is not possible to understand at first sight how they can be determined in the manner we have suggested. It is precisely the typical hysterical symptoms which fall into this class, such as hemi-anaesthesia, contraction of the field of vision, epileptiform convulsions, and so on. [S.E. 1893, 2: 5]

More generally, as Freud stressed, the products ("derivatives") of the dynamic unconscious that are thrust back into consciousness have a more or less *distant* topical connection to the primally repressed motif in which they presumably originated (S.E. 1915, 14: 149–150, 190–191). And the topical remoteness of these derivatives (symptoms, free associations, and phantasies) had been accentuated almost the moment Freud ceased his collaboration with Breuer: He soon demoted etiologically—to mere *precipitators* of neurosis—the adult repressions uncovered by his mentor's cathartic method (S.E. 1896, 3: 194–195). The childhood repressions that then supplanted adult ones as the primogenetic pathogens of hysteria were invariably sexual, thus being further removed thematically, no less than temporally. And we need only think of Freud's homosexual etiology of paranoia to be struck by the topical distance between the suspicions or delusions manifesting this affliction, and repressed homosexual yearnings, which are its putative pathogen.

Similarly, obsessively conscientious performance of religious rituals is held to derive etiologically from allaying portentous diffuse anxiety, generated by carnal temptation (S.E. 1907, 9: 124). Again, the topical "distortion" (S.E. 1915, 14: 149–150) deemed characteristic of much symptom-formation is also illustrated in Ferenczi's attribution (Jones 1938: 158) of phobic blindness to repressed castration anxiety. In short, an agent unconsciously effects a "compromise" when thwarted in the pursuit of an instinctual objective: He (she) settles for the substitutive satisfaction afforded by alternative conduct. And precisely when there is thematic *remove* from the original aim, this surrogate behavior would strike the agent as motivationally (causally) *unrelated* to it. All the same, due to the empathic appeal of thematic affinity, it has even been made a touchstone of the causal connection between psychic states: On Karl Jaspers's view (1973: 380–381), this connection is only a neuro-physiological *epiphenomenon*, unless there is thematic affinity.

So much then for the conceptual context in which Ricoeur has attempted to embed his *semiotic* construal of the various outcroppings of repressed ideation. On his conception, these derivatives can be assimilated to linguistic communications, such that the subject-matter of the clinical theory allegedly has the status of a veritable "semantics of desire". To gain perspective on the appraisal of this linguistic turn, let us first consider certain commonplace partial effects of purely physical states of affairs that, I submit, offer an instructive parallel—in relevant respects—to the symptoms of repressions in Freud's conative domain.

A foot-shaped configuration of the sand on a beach almost invariably originates causally from the incursion of a foot when a person or an animal walks on the beach. Hardly ever does such a beach formation result from the "chance" collocation of sand particles under the action of, say, some gust of wind. By thus being statistically linked to a prior physical interaction of the beach with a similarly shaped agency, the pedal sand structure qualifies *onto-*

logically as a *trace, mark,* or *imprint* of the past incursion (*cf.* Grünbaum 1974, chap. 9 for details). On the strength of this ontological status as a trace, the footprint can be said to derive *retrodictive significance for a human observer, who sees it:* The onlooker is *epistemically* entitled to *interpret* the sand formation as *attesting* to the prior ingress. And in this physical case, the geometric iso-morphism — "thematic affinity" — between a foot and the *mark* of its penetra-tion licenses a retrodictive *inference* of the latter occurrence.

If we were now to speak loosely, though more solemnly, we could say that the sandy shape has "meaning". But this may well becloud an otherwise lucid state of affairs: *Qua* having originated *causally* from a pedal invasion, the sand configuration is epistemically a *veridical indicator* of this inroad. But there is no danger of conceptual legerdemain via *linguistic* overtones, if we were now to use a *semantic metaphor* by declaring: The pedal sand form retrodictively "*bespeaks*" or "betokens" the incursion event. For not even someone with Ricoeur's semantic penchant would be tempted, in *this* context, to trade on this metaphoric locution so as to assimilate the epistemic status of the trace to the *semantic* one of a *linguistic* sign. The footprint is *not,* as such, a vehicle of communication: It is not a linguistic sign or symbol; it does *not* semanti-cally stand for, denote, designate, or refer to the past pedal incursion. When a language-user verbalizes the inference of this event, then it is the *utterance* of this retrodiction — *not* the trace licensing it! — which has *semantic* "mean-ing". Whereas those "symbols" that qualify as linguistic signs do have *inten-sion* (in the semantic sense) and *extension* (denotation), traces do not. And, as we shall see, what is true of traces also holds for neurotic *symptoms,* even though — when conceptualized psychoanalytically — they can also be called "sym-bols" in virtue of having the *conatively vicarious* function of affording substitu-tive gratifications or outlets.

Even though footprints and (mini-)neurotic symptoms each license alike the retrodictive inferences to their putative causes, there is of course one im-pressive difference between them: The occurrence of full-fledged or mini-neurotic symptoms presumably affords *vicarious conative* satisfaction for the repressed, thwarted yearnings to which they attest, whereas no kind of substitutive ful-fillment of desire is implicated in the formation of physical traces as such. But this difference is as *unavailing* to Ricoeur's notion of "semantics of desire" as it is obvious. For the conatively *vicarious* status that Freud's theory attributes, for example, to manifest dream content — a mini-neurotic symptom — hardly bestows the *semantic* function of a linguistic vehicle of communication on such fancied wish-fulfillment. As one might therefore expect, Freud himself con-trasted even the rather indefinite "ancient languages and scripts" with manifest dream contents: The former, he tells us, "are always, by whatever method and with whatever assistance, meant to be understood. But precisely this characteristic is absent in dreams. A dream does not want to say anything to anyone. It is not a vehicle for communication" (S.E. 1916, 15:231).

Ironically, Ricoeur (1970: 5) adduces none other than dreams as his center-piece for contriving a semantic role for symptoms by trading on their cona-tively vicarious nature. As he puts it: "Freud invites us to look to dreams themselves for the various relations between desire and language." In an at-tempt to buttress this faulty exegesis, Ricoeur (1970) makes three claims: (i) "It is not the dream as dreamed that can be interpreted, but rather the text of the [spoken] dream account" (p. 5), a contention which we already found to be insidiously misleading. (ii) He misassimilates the *latent* dream content to "another text that could be called the primitive speech of desire" (pp. 5–6), as if repressed wishes were verbal devices of communication. He then con-cludes (iii), "It is not desires as such that are placed at the center of the [psycho-] analysis, but rather their language" (p. 6). By such legerdemain, he is able to contrive his notion that the domain of psychoanalytic theory is a veritable "semantics of desire".

That the insinuations of this locution are ill-conceived becomes trans-parent the moment one takes a concrete case from Freud's conative realm. Thus, when a paranoiac gives verbal or non-verbal expression (e.g., by suspicious glances) to his (her) persecutory delusions, this distrustful behavior is good *evidence,* for a partisan of psychoanalysis, that the afflicted person is harboring repressed homosexual longings. But it is an arrant conceptual conflation to assimilate the vicarious paranoid outcroppings of these unconscious sexual feel-ings to speech acts that refer semantically to them. More generally, even when symptoms and other derivatives *are actually verbalized,* and even if they are the-matically cognate to their unconscious causes, they do not linguistically desig-nate the repressions that engender them, although they do manifest them!

Unencumbered by such considerations, Ricoeur (1970: 30) construes wish-fulfillment *not* as the achievement of a desired outcome, but rather à la Husserl as the fulfillment of a "signifying intention". If that were so, then Freud's theory of dreams would become a branch of the theory of descriptive seman-tics, which deals with natural — as distinct from formalized — languages. In-deed, as Shope (in press) has noted, Ricoeur endorses Lacan's obfuscating view that a symptom is like a language whose speech must be realized, whatever that is. If this characterization were appropriate for neurotic symptoms, why would it not also be applicable to psycho-*somatic* and even somatic symptoms?

For example, if a person afflicted by an as yet undiagnosed subdural

haematoma has severe headaches, why could one not extrapolate to this somatic symptom Ricoeur's referential account of all neurotic behavior? One
might then say that the headaches hint darkly at the patient's intra-cranial
pressure, because they are a kind of indirect language whose hidden reference
it is the task of *interpretation* to articulate. Hermeneutically "unenlightened"
neurologists refer to such interpretations as "diagnoses", just as physicists who
seek a theoretical identification of the particles producing particular cloud
chamber tracks, for example, do not see any point in speaking of their activity
as "deciphering track texts hermeneutically". For it would be insipid to foist
a semantic role on somatic symptoms by calling them "tell-tale", for example,
with a view to then seeing them as an indirect language. Yet Ricoeur insists
on contriving a linguistic construal for dreams as dreamt (1970: 15): "It must
be assumed . . . that dreams in themselves border on language, since they
can be told, analyzed, interpreted." But all manner of physical occurrences —
e.g., solar flares, barometric drops, and the productions by quasars or pulsars
— can be, and are, reported, *interpreted,* or analyzed, though not of course *psycho*-
analyzed.

In any case, it is spurious to assimilate to one another the following
two sets of relations: (i) the way in which the effect of a cause *manifests* it
and hence can serve *epistemically* as evidence for its operation and (ii) the manner in which a linguistic symbol represents its referent semantically or desig-
nates the latter's attributes. To be an effect *E* of a certain cause *C*, so that
E manifests *C* and is evidence for it, and to be *furthermore* a *conative surrogate,*
is plainly quite different from being any kind of linguistic representative: Paranoid *behavior* may well be a *vicarious outlet* for repressed homosexuality, but
in no case is it a verbal label for it! Thus as we saw, etiologically that behavior
is the afflicted person's attempt to cope with the anxieties generated by his
unconscious sexual urges, *not* his (her) attempt to *communicate* these yearnings
by means of persecutory delusions and behavior. Yet Ricoeur has left no stone
unturned to insinuate the assimilation of epistemic inferrability and conative
vicariousness, on the one hand, to semantic reference, on the other. And he
does so on the unavailing ground that analysts speak of symptoms as "symbols" when they are concerned to convey that symptoms qualify as *conative
surrogates qua* compromise formations. Once this logical state of affairs is articulated, the unilluminating and indeed seriously misleading character of Ricoeur's "semantics of desire" becomes all too apparent.

Robert Shope (1973; in press) has contributed a painstaking textual ex-
amination of Freud's several uses of the term "meaning", and of those passages
from Freud that Ricoeur adduced to support a semantic construal of symp-

toms. In a telling rebuttal of that exegesis, Shope successfully controverts a reading, according to which "Freud thought that mental phenomena such as dreams have meaning in the manner in which a language or speech signifies something" (1973:284). The manifest content of a dream "points to" the latent content, Shope rightly avers, *not* by designating it semantically, but "only in the sense in which a clue points toward something" (1973:285). As illustrations, he mentions (i) rock strata that serve as clues to the past for the geologist, "but not through any intentionality" (p. 302), (ii) physical symptoms that indicate the presence of a viral disease, and (iii) a rise in prices betokening a change in some other economic variable (p. 293). Hence Shope sums up lucidly (p. 294):

> Freud views the relation between these mental phenomena [dreams, symptoms, or parapraxes] and their meaning as similar to the relation between the symptoms of measles and its cause. They express the underlying states as effects manifest a cause. They are signs only in the sense that organic symptoms are signs of a diseased organism, or dark clouds signs of rain to come. Symptoms are signs to the investigator, and may arouse his expectations about finding their source, as dark clouds may arouse expectations of rain. If the investigator chooses, he may make these phenomena or representations of them into signs, in Ricoeur's sense, in the investigator's sign language. But to the patient symptoms do not yet designate or intend their underlying sources, any more than the darkness of a cloud designates or intends rain to the cloud or to any person before an observer devises a sign language. Rather, the symptoms of neurosis stand for their psychic sources in the sense of being stand-ins, that is, they appear in consciousness in place of the appearance of their hidden meaning; they are substitutes for it.

And in his more recent paper, Shope (in press) put the cogent upshot of his earlier study as follows:

> To say something is a stand *in,* i.e., something that occurs because a wish is not consciously admitted, or, in Ricoeur's words, something that occurs in place of "what the desire would say could it speak without restraint" (1970: 15), is not to say that the symptom stands *for* the (content of) the wish in the sense of referring to it.

Yet Ricoeur saw fit to declare (1970:359):

> Psychology is an observational science dealing with the facts of behavior; psychoanalysis is an exegetical science dealing with the relationships of meaning between substitute objects and the primordial (and lost) instinctual objects. The

two disciplines diverge from the very beginning, at the level of the initial notion of fact and of inference from facts.

5. Ricoeur's Disposition of "The Question of Proof in Freud's Theory"

We can now conclude our scrutiny of Ricoeur's chapter "The Question
of Proof in Freud's Writings" (1981, chap. 10). In its last pages (268–273),
he develops the import, as he sees it, of his earlier arguments for the answer
to the following query: "What sort of verification or falsification are the statements of psychoanalysis capable of?" But, as we saw, he speciously sets the
stage for his reply by the following stratagem: (i) He tendentiously mutilates
the purview of the psychoanalytic theoretical corpus by peremptorily deeming
it coextensive with "a work of speech with the patient" (1970:369), thereby
excluding even the analysand's nonverbal productions. (ii) He espouses a discredited observation-theory dichotomy for the natural sciences, and miscontrasts the analyst with the natural scientist epistemically ("the analyst does
not observe, he interprets"), whereupon he concludes that "there are no 'facts'
nor any observation of 'facts' in psychoanalysis but rather the interpretation
of a narrated history" (1974:186). (iii) He contrives a pseudo-dichotomy between psychoanalysis and "academic psychology", after misequating the latter
with a crudely conceived behaviorism and derives a spurious moral from the
thesis that "psychoanalysis deals with psychical reality and not with material
reality" (1981:251). (iv) He carries out a misassimilation of neurotic symptoms *qua* compromise-formations to *linguistic* representations of their hypothesized unconscious causes.

No wonder, therefore, that his treatment of the problem of validation
of psychoanalytic theory turns into a question-begging parody of what it ought
to be, as I shall now illustrate.

1. He offers nothing toward the validation of the *causal* hypotheses with
which, he acknowledged (1981:262–264, 269–270), Freud's clinical theory is
replete! For example, Ricoeur's first criterion of "the validation apt to confirm the truth claim belonging to the domain of psychoanalytic facts" is altogether *self*-validating and stale: "A good psychoanalytic explanation must be
coherent with the theory or, if one prefers, it must conform to Freud's psychoanalytic system" (1981:271).

His evasion of the corroboration of the causal efficacy of analytic treatment furnishes another instance. On the one hand, he tells us (1981:263) that

a psychoanalytic explanation offered to the patient "is itself [held to be] a causally relevant factor in the [therapeutic] work—the working through—of analysis". But when offering his third criterion of validation for Freudian theory, he remains altogether silent as to how—short of resort to the eschewed methods of the empirical sciences—he proposes to *demonstrate* therapeutic efficacy so as to fulfill the following demand: "A good pyschoanalytic explanation must . . . become a therapeutic factor of amelioration. . . . therapeutic success . . . constitutes in this way an autonomous criterion of validation" (1981: 272–273).

2. He offers a naive, if not smug, dismissal of the completely unsolved problem of epistemic contamination of the analysand's responses by suggestion, which he sees as arising from a "crude" objection (1981: 270):

> I shall leave aside the crude . . . objection, namely, that the analyst *suggests* to his patient that he accept the interpretation which verifies the theory. I am taking for granted the replies which Freud opposes to this accusation of suggestibility. They are worth what the measures taken at the level of the professional code and the analytic technique itself against the suspicion of suggestion are worth.

As Part I of my (1984) shows in detail, Freud valiantly, yet unsuccessfully, struggled throughout his life to hold just this epistemic objection at bay, after deeming it "uncommonly interesting" (S.E. 1917, 16: 447). And, Freud himself (S.E. 1917, 16: Lectures #27 and #28) appreciated all too keenly that even if the analyst does his best to forego overt or covert suggestion, there are myriad ways in which he can unconsciously but persuasively mold the analysand's convictions and engender a compliant pseudo-corroboration. Hence it plainly won't do to adduce the analyst's professional integrity and avowed intent not to abuse his suggestive influence as an adequate safeguard against the elicitation of *spurious confirmations* from the patient.

3. Ricoeur concludes by invoking the probatively synergistic, cumulative character of his proposed criteria of validation. He does concede that this purported "proof apparatus of psychoanalysis . . . is . . . highly problematical". Yet the fact remains that we are left completely in the dark as to how any *one* of his criteria of validation can be met at all within the avowed *confines* of (i) the purely verbal *intra*-clinical "facts" countenanced by him as constituting the purview of Freud's theory and (ii) the renunciation of natural science modes of causal validation in favor of purely hermeneutic devices of some sort. Hence it would seem that, under these restrictions, his criteria are, collectively no less than severally, quite unhelpful.

It emerges that Ricoeur's hermeneutic construal of psychoanalysis is anything but cogent, logically or even exegetically. Yet R. S. Steele (1984) pays tribute to him as follows:

> Ricoeur provides Freud a haven from his scientific critics by granting them their criticism. He agrees that "psychoanalysis is not an observational science," but he uses this admission to counter behaviorist, experimental, and logico-empiricist attacks on psychoanalysis.

6. Concluding Remarks

The "reading of intentionality" or the discernment of "latent meaning" in psychoanalysis, which hermeneuticians are wont to adduce, is none other than the use of Freud's clinical postulates to interpret human thought, affect, and behavior theoretically in individual cases by inferring its hypothesized unconscious *causes*. For particular psychoanalytic interpretations *articulate the psychic contents* to which the general clinical hypotheses attribute particular symptoms, slips, and dreams—repressed contents whose manifestation was purportedly distorted by the defensive operations of censorship, displacement, condensation, etc.

But physicists too "read" phenomena in the sense of interpreting them theoretically by hypothesizing explanatory causes for them. And such physical cosmogonies as the big bang model of the universe are partly historical or narrative, as is historical geophysics. Similarly, Darwinians "read" biogeographical distributions by offering explanatory historical narratives for them. Furthermore, it is elementary and commonplace that the generic *identity* of the phenomena interpreted by the psychoanalyst is ideational, and therefore differs in *identity*, at least *prima facie*, from the phenomena "read" by the physicist. Hence, in addition to being potentially misleading, it is just turgid to inflate this utter banality by designating the psychoanalyst's *interpretive* stock-in-trade as being "hermeneutic". Nor is there any need to label the *meta*-interpretative *philosophy* of psychoanalysis as "hermeneutic", any more than it is illuminating to so designate, say, the meta-interpretative philosophy of physics.

True, it was Dilthey rather than Ricoeur, Habermas, Klein, et al. who grafted the term "hermeneutics" from its original *philological* context (biblical exegesis) onto psychology. But whatever the verdict on its appropriateness to Dilthey's ideographic, anti-nomothetic conception of psychology, I have argued that its extrapolation to Freud's nomothetic clinical theory begets conceptual mischief. For its *non-trivial* applicability to this body of hypotheses is predicated on the philosophical theses of the latter-day hermeneuticians, which have turned out to be quite alien to psychoanalytic theory. For this reason alone, it would appear ill-advised to elevate the hermeneutic construal of psychoanalysis to the status of a paradigm for the human sciences.

Notes

1. All references to Freud's writings in English will be to the *Standard Edition of the Complete Psychological Works of Sigmund Freud,* translated by J. Strachey et. al. London: Hogarth Press, 24 volumes. Each reference will employ the acronym "S.E.", followed by the year of first appearance, the volume number, and the page(s).

2. Barbara von Eckardt (1984) has given a lucid discussion of my arguments against Habermas in a helpful digest of my philosophy of psychoanalysis.

References

Brenner, C. 1982. *The Mind in Conflict.* New York: International Universities Press.

Freud, S. 1954. *The Origins of Psychoanalysis.* New York: Basic Books.

———. 1966–1974. *Standard Edition of the Complete Psychological Works of Sigmund Freud,* 24 volumes, translated by J. Strachey et al. London: Hogarth Press.

Gauld, A., and Shotter, J. 1977. *Human Action and Psychological Investigation.* London: Routledge & Kegan Paul.

Grünbaum, A. 1974. *Philosophical Problems of Space and Time.* 2d enlarged ed. Dordrecht and Boston: D. Reidel.

———. 1983. "Freud's Theory: The Perspective of a Philosopher of Science". 1982 Presidential Address to the American Philosophical Association (Eastern Division). *Proceedings and Addresses of the American Philosophical Association* 57: 5–31.

———. 1984. *The Foundations of Psychoanalysis: A Philosophical Critique.* Berkeley: University of California Press.

Habermas, J. 1970. *Zur Logik der Sozialwissenschaften.* Frankfurt: Suhrkamp Verlag.

———. 1971. *Knowledge and Human Interests,* translated by J. J. Shapiro. Boston: Beacon Press.

———. 1973. *Theory and Practice.* Boston: Beacon Press.

Jones, E. 1938. "The Theory of Symbolism". Chapter 6 in his *Papers on Psycho-Analysis.* London: Bailliere, Tindall & Co.

———. 1959. *Editorial Preface to S. Freud, Collected Papers.* Vol. 1. New York: Basic Books.

Klein, G. S. 1976. *Psychoanalytic Theory.* New York: International Universities Press.

Masson, J. M. 1984. *The Assault on Truth.* New York: Farrar, Straus & Giroux.

Maxwell, G., and Maxwell, M. L. 1972. "In the Beginning: The Word or the Deed". A review of Ricoeur's *Freud and Philosophy. Contemporary Psychology* 17: 519–522.

Moore, M. M. 1983. "The Nature of Psychoanalytic Explanation", in L. Laudan, ed., *Mind and Medicine: Explanation and Evaluation in Psychiatry and the Biomedical Sciences.* Pittsburgh Series in the Philosophy and History of Science, vol. 8. Berkeley, Los Angeles, London: University of California Press.

Ricoeur, P. 1970. *Freud and Philosophy.* New Haven: Yale University Press.

————. 1974. *The Conflict of Interpretations,* Don Ihde, ed. Evanston, Il.: Northwestern University Press.

————. 1981. *Hermeneutics and the Human Sciences,* translated by J. B. Thompson. New York: Cambridge University Press.

Schafer, R. 1976. *A New Language for Psychoanalysis.* New Haven: Yale University Press.

Shope, R. K. 1973. "Freud's Concepts of Meaning". *Psychoanalysis and Contemporary Science* 2: 276–303.

————. In press. "The Significance of Freud for Modern Philosophy of Mind", to appear in *Contemporary Philosophy,* vol. 4, *Philosophy of Mind,* edited by G. Floistad. Boston: Nijhoff.

Steele, R. S. 1979. "Psychoanalysis and Hermeneutics". *International Review of Psychoanalysis* 6: 389–411.

————. 1984. [in press]. "Paul Ricoeur & Hermeneutics", forthcoming in J. Reppen, ed., *Beyond Freud: A Study of Modern Psychoanalytic Theorists.* Hillsdale, N. J.: The Analytic Press.

von Eckardt, B. 1984 [in press]. "Adolf Grünbaum and Psychoanalytic Epistemology", in J. Reppen, ed., *Beyond Freud: A Study of Modern Psychoanalytic Theorists.* Hillsdale, N.J.: The Analytic Press.

Explaining the Success of Science: Beyond Epistemic Realism and Relativism[1]

LARRY LAUDAN

1. Introduction

Throughout the first half of the eighteenth century, scientific opinion concerning the structure of the cosmos was deeply polarized; numerous "systems of the world" found their advocates among prominent natural philosophers, but the two leading rival systems were those of Descartes and of Newton. Cartesian physics held sway in France and on much of the rest of the continent; Newton's reigned in England. The young Voltaire journeyed from Paris to London in the spring of 1727. He was confused by the contrasting world-views he found. With an acute case of culture shock, he wrote to a friend back home:

> A Frenchman who arrives in London finds a great shift in scientific opinion that makes the mind weary. He left the world full; he finds it empty. At Paris you see the universe composed of tiny vortices of subtle matter; in London we see nothing of the kind. . . . With the Cartesians, all change is explained by collisions between bodies, which we don't understand very well; with the Newtonians it is done by an attraction which is even more obscure. In Paris you fancy the earth's shape like a round melon; at London it is flattened on the two sides.[2]

One gets the same dizzying and disorienting feeling in our time if one moves between circles of philosophers and sociologists of science. Many, perhaps most, philosophers in the analytic tradition (and especially philosophers of science), take it for granted that science is, at least in its essentials, largely true and substantially correct. These philosophers argue that, especially in the "mature" and well-developed parts of the physical sciences, scientists have come very close to discerning the way the world *really* is. Our theories about such matters are, they say, highly verisimilar. Even where science turns out not to be strictly true, most philosophers (present writer included) are still apt

83

to consider science as our best exemplification of rationality and cognitive progress—our best guess as to how things stand.

Sociologists, by contrast, especially sociologists of knowledge, tend to see science differently. Many of them regard scientific theory, like science itself, simply as a social construct, a set of conventions which Western culture since 1700 has used for conceptualizing experience, but which has no particular purchase on reality. Every culture, they point out, has its myths and its sacred beliefs; we happen to call ours by the name 'science'; but those beliefs are no better, no more secure, objective, or rationally grounded than the guiding ideologies of other cultures. These two points of view are known as 'realism' and 'relativism' respectively. The pair of them and the injustices that each does to an understanding of science will form the foci of this paper.[3]

But before I turn to that task, one crucial qualification is in order concerning the compass of this essay. Both realism and relativism have received numerous (and often conflicting) formulations by a wide variety of writers. While there may be fewer realisms and relativisms than there are realists and relativists respectively, it is a close call. There are many species of both these groups which I shall not be discussing in this essay. Perhaps the best way of locating my concerns is to say that I will be grappling with the *specifically epistemic* formulations of realism and relativism. Equally familiar to most readers will be various *ontological* versions of both realism and relativism. I must emphasize at the outset that the latter, metaphysical theories are *not* the targets of my criticism. My preoccupation in this essay with epistemic and methodological matters should explain why I shall have little to say about many varieties of relativism (e.g., Quine's ontological relativity) which might otherwise be expected to occupy center stage in a critique of relativism.

Epistemic realism or, to be more precise, 'scientific realism', has been the reigning orthodoxy among philosophers of science for almost a generation. Philosophers as diverse in orientation as Popper, Grünbaum, McMullin, Sellars, Reichenbach, and Putnam have espoused one or other version of it. Its rival, epistemic or cognitive relativism, has found occasional philosophical advocates (e.g., Feyerabend, the later Wittgenstein, Hesse and Rorty) but cognitive relativism—at least in that variant of it which I shall treat here—is associated primarily with work in the sociology of knowledge. Mannheim, Durkheim and Kuhn have developed what are probably the three most familiar versions of this species of relativism.[4]

Both realism and relativism are theories of knowledge in the broadest sense, and both have complex ramifications for our understanding of science. To put it briefly, the realist insists that science, in the course of its development through time, provides us with an ever more accurate, an ever more nearly true, representation of the natural order. Scientific theories, if not strictly true, are nearly so; and later scientific theories are closer to the truth than earlier ones. More than that, the realist typically asserts that science or the scientific method represents the *only* (or, more weakly, the *most*) effective instrument for discovering truths about the world. The relativist, by contrast, characteristically eschews notions of truth and falsity, focussing rather on the specific and local features which shape (and, in his view, inevitably distort) the scientific image of the world. The relativist would have us believe that science is but one among indefinitely many ways in which man might represent the world; in his view, it has no special claim to validity or veracity. If we lived in a different place and time, says the relativist, we would have a fundamentally different vision of the natural order. Still worse, there is— the relativist maintains—no neutral point on which we can stand to adjudicate impartially the rival claims of these contrasting images of the world, the scientific and the nonscientific. Because we ourselves are products of a scientific culture, we cannot step outside the presuppositions of that culture to compare the legitimacy of its claims with those of nonscientific cultures. Where the realist sees the history of science as a triumphal march ever closer to the truth, the "cutting edge of objectivity" (in Gillispie's apt if notorious phrase), the epistemic relativist sees nothing more than a succession of rival and mutually incompatible representations, each reflecting various subjective and transitory interests. Where the realist sees progress in the history of science, the relativist sees only change. The realist believes that science comes as close to truth and objectivity as is humanly possible; the relativist fears that he is probably right! But they draw very different conclusions from this one point of consensus.

There is nothing especially new about this polarity. Struggles between realist and relativist perspectives span the entire history of epistemology. Precisely because of their age-old opposition, there is a tendency to see these doctrines as mutually exhaustive rivals. Any weakness in relativism (e.g., its allegedly self-indicting character) is translated into an argument for realism; while any flaw in realism (e.g., the unsatisfactory status of realist semantics) comes to be widely regarded as evidence for relativism.

A number of considerations move me to take strong exception to the view that these two doctrines more or less exhaust the range of alternatives open to us. Both seem to me to be fundamentally flawed and open to anomalies which are beyond their resources to grapple with. But more important, each fails to resolve one of the most central conceptual questions about science. In a nutshell, that question is simply: why does science work so well? In what follows, I shall seek to show:

1). that the realist recognizes the importance of this question but fails to answer it;

2). that the relativist is scarcely prepared to grant the legitimacy of the question, let alone to answer it;

3). that the question can be interestingly answered, provided that we are prepared to lay aside some of the core assumptions associated with both realism and relativism.

2. Establishing the Phenomenon: The Success of Science

As every student of scientific controversy understands, one man's fact is another man's fiction. (Recall, if you have any doubts, the difference between London and Paris in 1727). Nowhere is this difference in "perception" more marked than with respect to the question of the success of science. Many of us incline to the view that nothing could be more obvious than the fact that science is a successful and effective knowledge-gathering enterprise. We may be unsure how to account for that success and our efforts to *characterize* it precisely have not been very illuminating (witness the failure of the many inductive logics and theories of confirmation); but our intuition remains unshaken that science does what it does very well indeed. It is quite another matter, however, where sceptics and relativists are concerned. It is not clear whether they positively deny that science is successful; in general, they simply do not reckon the achievements of science to be something which they are called upon to explain. Insofar as they deal with the phenomenon at all, it is to point out that the "success" attributed to science is of an ambiguous and amorphous sort. "Successful according to whom? and for what purposes?" they ask. "Successful compared to what?" "Successful by which standards?" To put the most sympathetic gloss I can on the relativists' failure to grapple with the problem of the success of science, I would say that relativists are inclined to withhold judgment on the claim that science is successful for these reasons (among others): (1) a belief that, until the notion of success is spelled out with some care, the concept is too unclear to be worthy of systematic analysis, let alone explanation; and (2) a lingering suspicion that 'success' is an evaluative rather than a descriptive term which should play no role in an empirical and naturalistically-based sociology of knowledge.[5]

What I want to do in this section is to meet that challenge by describing some notion(s) of success which should allow both relativists and realists to grant that science is, more than occasionally, successful. We can then proceed to explore the resources of realism and relativism respectively for accounting for that success.

We must begin by freeing 'success' of some of its more normative and judgmental overtones. As I shall be using the term, judgments of success in an activity imply no endorsement of that activity. There can be successful bank robbers, rapists, military campaigns, or scientific theories. One may or may not regard science and technology as forces for good; but no such evaluation is presupposed or implied by my claim that science is a successful activity. In the most general sense of the term, success in any activity always has to do with relations between ends and means and, more specifically, between aims and actions. To say that an activity is successful is simply to say that it promotes the ends of (at least some of) those engaged in it (or, and this is an important codicil, of those judging it to be successful). Just as we say before the fact that an action is rational if the actor has good reason to believe that it will achieve his goals, so do we say *post hoc* that an action is successful just insofar as it actually furthers some agent's goals.[6] Putting it this way makes it clear that success is a *relational* concept. Because agents' goals can differ, one and the same action may be unsuccessful or successful, depending upon the goals in question.

Of even greater importance is the fact that success, so conceived, is not a valuational or a normative concept. To claim that a certain action was successful is to make a contingent, empirical claim about the relation of that action and its outcomes to certain goal states. Claims about success are thus (at least in principle) as factual and as testable as any other sort of empirical claim about the world. (Although one must concede the fairness of the relativist's charge that, in practice, 'success' is too often treated as a primitive term.)[7]

Accordingly, the thesis that science is successful (or unsuccessful) amounts to the empirical assertion that the actions of scientists have in fact brought about or otherwise promoted (or failed to promote) certain goals or aims. But with respect specifically to which goals is science to be judged successful or unsuccessful? This question is both simpler and more complex than it might first appear. It is simpler because, unlike judgments about the rationality of an action, judgments of success do not require a scrutinizing of an agent's aims or motives. We can ask whether an agent's actions in fact brought about certain outcomes, quite independently of whether those outcomes were the ones which the agent intended to achieve. Just so long as we make it clear what outcomes we regard as constituting 'success', we can happily make determinations of success or failure without skating on the comparatively thin ice of attributions of intentionality to agents. Of course, we will often be interested to know whether an action is successful specifically with respect to (what we take to be) the goals of the agent who embarked on the action. But making determinations of 'success' parasitic on the agent's goals is not necessary, and in this particular case is almost certainly not desirable.[8] Through time, scientists have had a highly heterogeneous set of cognitive aims or goals.[9] There almost certainly is no such thing as *the* aims of the scientific community, any more than any other large and diverse group has universally shared and univocal aims. When we say that science has been successful (at least those of us who are prepared to venture such a conjecture), we do not usually even bother to engage in a detailed analysis of the goals or aims of all or most of the actors who have constituted the scientific community. Rather, we typically *impute* certain goals to a highly idealized caricature of the scientist (or, even more abstractly, to science as an institution) and then ascertain whether science has achieved those goals. It seems to me that it would be a more forthright and intellectually honest approach to admit that, at least for these purposes, we are not trying to ascertain whether science has managed to achieve what, as a matter of fact, working scientists have always or invariably been trying to achieve. That is certainly an interesting question, but it is not the most important question for epistemic or methodological purposes. Rather, we should say straight out that, when we are judging whether science is or has been successful, we are going to be making determinations of success with respect to the ability of science to achieve certain cognitive attributes which *we* find especially interesting. As long as we acknowledge what we are doing, we can avoid vexed questions about intentionality, incompatible goals, shifting explanatory ideals, and the host of other difficulties which confound efforts to ascertain what the aims of science have actually been through history. By taking this route, we can make our task a good deal easier than it otherwise would be.

What makes the task rather more complicated than one might expect is the necessity of spelling out clearly and precisely exactly what criterion of success we are utilizing. Since virtually any action will have some consequences or other, some outcome or other, it is always possible after the fact to find some set of descriptions under which any particular action can be made to appear to be successful. Such an approach would obviously trivialize the undertaking. Accordingly, we need to find an interesting, unusual, and demanding set of outcomes, with respect to which we will proceed to make judgments of success or failure. (There is obviously no unique set of that sort.) But there is one set of cognitive outcomes which has interested epistemologists and philosophers of science for a long time. These goal states concern themselves with certain interesting epistemic and pragmatic attributes. Consider a typical list of some of those aims:

a). to acquire *predictive control* over those parts of one's experience of the world which seem especially chaotic and disordered;

b). to acquire *manipulative control* over portions of one's experience so as to be able to intervene in the usual order of events so as to modify that order in particular respects;

c). to increase the *precision* of the parameters which feature as initial and boundary conditions in our explanations of natural phenomena;

d). to integrate and *simplify* the various components of our picture of the world, reducing them where possible to a common set of explanatory principles.

If we define cognitive 'success' along these lines, then it seems uncontroversial to say that portions of the history of science in the last 300 years have been a striking success story. For instance, we are now in a position to predict a much broader range of phenomena than we were in 1700. We can intervene in the natural order (e.g., with respect to the course of many diseases) so as to make things go more to our liking far more effectively than we could formerly. Our instruments for measuring various variables and constants are incomparably more precise than they once were. (Consider, for instance, the refinements in the last two hundred years of determinations of the velocity of light.) Finally, even if the ultimate unification of science still eludes us, it is quite clear that we can now explain a more diverse set of phenomena in terms of a smaller number of general principles than our forebears could.

In saying that science has been successful in this cognitive sense, I am certainly not claiming that science has managed to achieve all the goals of all of its practitioners. Nor am I making any judgment about whether, all things considered, science is worthy or admirable. At least for purposes of this analysis, we need make no judgment about the moral or social value of the sorts of outcomes which science has achieved. I am here simply noting certain facts, and very striking facts they are, about the diachronic development of science. Because these facts are so striking, because there was no reason *a priori* to expect man to be able to achieve such cognitive feats, because no undertaking can guarantee success *of this particular sort,* we are confronted with a genuine problem: why is science so successful? What is it about the manner in which scientists formulate and test their theories which makes this sort of success possible? [10]

Any account of science which fails to answer, or even to address, such questions is (to put the criticism in its mildest form) fundamentally *incomplete*. Whatever one's disciplinary or philosophical orientation, one cannot pretend to be accounting for, or explaining, science in a comprehensive manner unless one has an answer to such questions as these. As we shall see, such is the sorry state of both realism and relativism.

3. Realism and Success

By and large, scientific realists have recognized the pivotal importance of the problem of explaining the cognitive success of science. Indeed, especially in the last few years, numerous realists have claimed that one of the chief arguments in favor of realism is precisely that it, allegedly unique among rival epistemologies, can explain why science is successful. Hilary Putnam, for instance, asserts that "the positive argument for realism is that it is the only philosophy that doesn't make the success of science a miracle".[11] McMullin, Newton-Smith, Boyd, and Niiniluoto have made similar claims on behalf of realist epistemology.[12] In this section, I want to examine briefly the claims of contemporary realism to be able to explain the success of science.

Realists argue that scientific theories, at least in such "mature sciences" as physics, are approximately true and that the central terms in such theories genuinely refer to objects in the physical world. They go on to insist that the approximate truthlikeness of our theories (and the related authenticity of reference exhibited by their central concepts) explains why science works as well as it does. Our theories are successful, the realist maintains, precisely because they come close to representing things as they really are.

What the realist is trading on here is the perfectly sound intuition that *if* our theories were true unqualifiedly, then all their consequences would likewise be true; and if all those consequences were true, we would indeed expect theories to exhibit just that sort of predictive accuracy and reliability which (on the account mentioned earlier) constitutes "the cognitive success of science".[13] Sadly, the realist is enough of a "realist" (in the hard-headed, ordinary language sense of that term), to recognize that the relevant antecedent conditions in this intuition are unsatisfied. We have overwhelmingly good reasons to suspect that our theories about the world, even our best-tested ones, are not true *simpliciter*. Yet the realist still wants to cash in on the hunch that the "truthlikeness" of our theories is responsible for their success. Accordingly, the realist maintains that, even if our best theories are only approximately true, or nearly true, then he is in a position to explain the success of science. The core idea here is that an approximately true theory will have consequences *most* of which are true, or at least which are close to the truth.

As I have shown elsewher‍ in detail,[14] this argument is fundamentally flawed. There is as yet no coherent sense of 'approximate truth' whicḷ entails that an approximately true theory will be uniformly successful in *any* of the senses sketched above. We can put the point more strongly: there is as yet no semantic account of truthlikeness which entails that a theory, all of whose central explanatory claims are approximately true, will be any more successful than a theory whose central explanatory claims are wildly inaccurate. It is entirely conceivable, for instance, that a theory might be approximately true, in any explored sense of the term, and still be massively inaccurate in those domains where it can be tested. Indeed, on the best-articulated sense of truthlikeness (namely Popper's theory of verisimilitude), it can be shown that a theory may have a high "truth content" yet have all its observable consequences false. To make a long story short, it can be shown that—because there is no reason to believe that approximately true theories need be (or are even likely to be) empirically successful—the near-truthlike status of theories—even assuming we had an epistemology which would warrant such attributions of truthlikeness—cannot be invoked to explain their pragmatic success.

But the situation is even gloomier than this for the realist. As even a brief glance at the history of science will show, there are many theories which have been highly successful for long periods of time (e.g., theories postulating spontaneous generation or the aether) which clearly have not been approximately true in terms of the deep-structure claims they have made about the world. Thus, Newtonian optics—which predicted a wide variety of phenomena, and which inspired the construction of a plethora of successful instruments and measuring devices—was committed to a basic ontology of light which (so we now believe) is desperately wide of the mark. Because there seems to be nothing in the world which even approximately corresponds to Newtonian light corpuscles, it is clear that Newton's theory was not, indeed could not have been, approximately true. (Assuming that, as I have argued elsewhere,[15] the scientific realist's notion of approximate truthlikeness presupposes genuineness of reference.) How then is the realist to account for the success of Newtonian optics? Even assuming that the realist could show that an approximately true theory would be successful (which he cannot), how can he explain the success of a theory, like Newton's, which is—by his lights—not even close to the truth? The same goes in spades for most theories in the history of science. Because they have been based on what we now believe to be fundamentally mistaken theoretical models and structures, the realist cannot plausibly hope to explain the empirical success such theories enjoyed in terms of the truthlikeness of their constituent theoretical claims. So it appears that the realist is in a bind. Theories which are approximately true need not be successful; many theories which have been strikingly successful are evidently not approximately true. Under such circumstances, truthlikeness is a decidedly unpromising explanans for empirical success.

But the problems facing the realist go even deeper than this. There is a crucial ambiguity in the problem of scientific success, an ambiguity which highlights another weakness in the realist approach to that problem. When we ask why scientific theories work so well, we might be asking (and the realist response assumes as much) to be told what semantic features theories possess in virtue of which they have such an impressive range of true consequences. Alternatively, when we ask why science is successful, we might be asking an epistemic and methodological question about the selection procedures which scientists use for picking out theories with such impressive credentials. If, as I suspect, it is generally the latter which we are driving at, then the appropriate response to that problem will address itself to the probative and evaluative procedures which scientists use for identifying those theories which are likely to be reliable. And insofar as that is what is at stake, the realist response becomes even less availing than it already appears to be; for the realist's "explanation" of success (viz., theories work because they are true or nearly true) sheds no light whatever on how scientists come by these putatively true or truthlike theories. Because the realist makes no reference to the methods of investigation and warranting which scientists use for selecting their theories, he must leave that side of the question unaddressed. That side of the problem of success remains a mystery on realist principles, even if the realist can get his theory semantics in order.

I will conclude this section by offering several caveats. I have asserted here neither that realist epistemology is wholly irrelevant to the explanation of scientific success nor that realism's failure to explain the success of science is a disproof of realism. (Although that failure does raise serious questions about whether realism has any empirical content.) What I have insisted is that current realist approaches to this issue provide little more than pseudo-explanations of the success of science. Beyond that, I have suggested that realists have largely missed the point about the success of science, for they have failed to see that what is chiefly called for is an epistemic analysis of the methods of theory testing rather than an account of theory semantics.

4. Relativism and Success

There are numerous variants of relativism (cultural, historical, and epistemological among others). What most seem to have in common is a conviction that no method of inquiry can claim special or privileged status. Different cultures, different societies, different epochs will exhibit conflicting views about the appropriate ways of authenticating beliefs. Confronted by these differences, the relativist insists on remaining agnostic about the respective merits of different methodological and evaluative strategies for testing claims about the world. More than that, the relativist generally denies in principle that there can be any way of showing one doxastic or belief-forming policy to be superior to another, or one set of methods to be objectively preferable to another.[16]

The relativist circumvents the problem of success by writing it off his explanatory agenda. As he sees it, his task is the descriptive and explanatory one of explaining why agents believe what they do. He will thus quite happily offer us an explanation for why a particular scientist or group of scientists believes that their theories are successful. But what he is reluctant to confront head-on, in part because he mistakenly imagines it to be a purely normative and philosophical puzzle, is the question why certain theories or beliefs are, in fact, successful.

Indeed, many latter-day relativists explicitly repudiate any effort to acknowledge that certain systems of belief have been more successful than others. David Bloor, Barry Barnes, and a host of other sociologists of knowledge have argued for such agnosticism;[17] so too have such philosophers as Paul Feyerabend. My aim in this section is to explore the relations between relativism, so understood, and the problem of the success of science.

As I said earlier, this reluctance to grapple with the success of science is due, in part, to the relativists' failure to recognize that 'success' can be characterised in a thoroughly descriptive rather than an evaluative way. Equally, they have been understandably skeptical about approaches which assume that all human agents have the same cognitive goals. As the relativist sees it, terms like 'success' and 'failure' smack of cultural chauvinism because they seem to suggest that all human actors have the same aims. But if one takes seriously the arguments offered above, it becomes clear that the claim that a certain piece of science is successful relative to certain aims is not tantamount to the claim that all agents (or even all scientists) have the same aims. It is simply the empirical claim that the developments in science have in fact promoted certain aims. Since relativists do grant that there are some features of science which can be treated as data to be explained or accounted for, the relativist who refuses to countenance the success of science (so understood) must explain why the success of science is any less a fact about science than, say, that science is a social activity or that scientists use certain mechanisms for generating consensus.

But the relativist's uneasiness about the seemingly judgmental aspects of 'success' is only a part of the story. I submit that a second factor which encourages the relativist to finessse the issue of explaining why science is successful is his realization that his epistemology lacks the explanatory resources to give a plausible analysis of that success. Indeed, the fact that science is so successful constitutes a powerful anomaly to relativism; not because the success of science refutes relativism, but rather because it points up its descriptive and explanatory incompleteness as an empirical theory of the scientific enterprise.[18] Like some Victorians, who hoped venereal disease would go away if no one mentioned it, the relativist apparently thinks that aloof indifference to success is the preferred vehicle for wishing it away. Let me explain why I make this charge of incompleteness against relativism.

Consider one specific part of the success of science, its predictive ability. With respect to many sorts of phenomena, it is quite clear that science puts us in a position to anticipate what the world will do next with a rather higher reliability than can many systems of belief commonly regarded as nonscientific.[19] I submit that we have enormous support for the success thesis in that form. Let me particularize it still further in terms of a familiar example. In virtually every society which cultivates its own food, there is (what Habermas has called) a "technical interest" in anticipating when floods will occur, in judging their intensity, and in taking appropriate action to control the damage they wreak. Every agrarian and post-agrarian society has means for anticipating when and where major rivers will overflow their banks. Let us suppose, for the sake of argument, that modern Western scientific techniques yield predictions which are both more detailed and more accurate with respect to the phenomena of river flooding. If, as I believe, this greater accuracy could be convincingly established (even to those who were not products of Western culture), then we would be confronted with a situation where different cultures have a common technical interest in predicting a certain sort of phenomena and where scientific culture, by the standards of all concerned, yields more accurate predictions.[20]

What is the relativist to say about such a case? Well, what most of us non-relativists would say is that science has available certain methods of theory selection and theory testing which, over the long run, tend to pick out theories of high reliability. We might go on to point out precisely why one would expect methods of the sort scientists use to yield fairly reliable theories.[21] In the case in hand, we could contrast the methods of theory evaluation used in scientific predictions of rainfall and water run-off quantities with those rule-of-thumb methods utilized for generating predictions about flooding in non-scientific societies. I expect we would be able to show that our methods of theory selection are more robust than those of other cultures, thereby explaining why our theories about flooding were more efficacious in producing reliable predictions than the theories used by other societies.[22]

But, of course, such explanatory manoeuvres are not open to the relativist, for he denies that any methods can objectively be said to be better than any others.[23] Precisely because he makes that denial, he is in no position to cite the superior methods of science as the explanation of the greater predictive success of science compared to other forms of knowledge.[24] Since the relativist cannot explain the predictive superiority of science by invoking the greater rigor or robustness of its methods, what is he to say? Well, he might say that it is just a large cosmic coincidence that science is so successful; that the success of science reveals nothing about, and owes nothing to, the specific methods of inquiry used in science. But this is not to explain the success of science; it is, rather, to renounce any effort to account for that success.

I have suggested two causes for the relativist's reluctance to grapple with scientific success; one concerned the relativist's explanatory agenda (i.e., his uneasiness about "evaluative" concepts); the other, his limited explanatory resources. But I think that we must probe still further before we fully understand why many relativists are so reluctant to acknowledge the success of science as a datum to be explained. For more than a decade, relativist sociologists have been committed to the idea that the same sort of institutional analysis which they offer for other social structures and systems of belief (e.g., religion or the kinship system) can be applied indifferently to science. In their view, science is simply one among many institutions for the formation and perpetuation of beliefs. These new-wave sociologists have thereby sought to distance themselves from the older sociological tradition (e.g., associated with Robert Merton, among others) which tried to establish the cultural or sociological *uniqueness* of science as an institution. As soon as the relativist grants that science has been cognitively more successful than many other belief-building enterprises, then he can no longer argue for a monolithic or unitary account of cognitive practices. Put differently, the latter-day sociologist of knowledge wants to reduce the sociology of knowledge to the sociology of belief, and to conjoin that reduction with the thesis that the problem of explaining belief-

or consensus-maintenance is to be handled in a unitary fashion across all institutions and cultures. To maintain this homogeneity thesis, the relativist must either deny that science has been successful, or insist that it has been no more successful than any other system for the generation of action-related beliefs (or, finally, hold that the success of science is fortuitous). In either case, the relativist's denial that science is *sui generis* disposes him to deny that the success of science is a datum requiring special explanation.

I have ventured into this lengthy digression about the causes that have evidently pushed the relativist in the direction of ignoring the problem of the success of science only because, when we find a group of thinkers denying what most of us take to be obvious, we have to cast about for some explanation of their apparently pathological behavior. But however far one goes in trying to understand why relativists might want to avoid grappling with the problem of the success of science, the fact remains that it is a phenomenon which they have left unexplained. To that extent, relativism is radically incomplete as an explanatory theory about science.

If there is any plausibility in the arguments of these last two sections, we seem in fairly dire straits. Neither of the major epistemologies of our time seems to show much promise of handling one of the core intellectual issues of our time. The better part of valor might suggest that the problem of the success of science is simply intractable, a problem well beyond our limited explanatory capacities. It is, after all, conceivable, that—as Karl Popper once suggested[25]—why science works is just an insoluble problem which is best left well enough alone. But before we acquiesce too quickly in that view, and before the relativists and realists imagine that they are off the hook (for who can reasonably be expected to solve an insoluble mystery?), it is worth exploring briefly whether we really are in such a desperate position so far as explaining why science succeeds.

5. Accounting for the Success of Science

This is, of course, a tall order, and I have no intention of offering here a perfectly general solution to the problem. What I will do is to take one or two typical, if slightly idealized, cases of scientific success and offer a story about them, a story which will make it plausible why science works well in those circumstances. A different story would have to be told about other cases. But if the tale I have to tell is at all convincing, it will be easy to see how it could be adapted to a wide range of other situations.

But before I offer my narrative, I need to make some important disclaimers. It has often been assumed that the demand for an explanation for the success of science (i.e., an account of why science "works" so well) is really just a re-formulation of the hoary old problem of induction. And it is true that a solution to the riddle of induction, assuming one could be had, might well give us a solution to the problem of success. After all, if we could show under what circumstances it was reasonable to assume that unobserved instances of a generalization or theory will resemble observed ones, we would have shown the reasonableness of (at least some) inductive methods. I do not have a solution to the problem of induction; come to that, I do not regard it a particularly interesting problem in this form. The point I want to make here is that the problem of the success of science can be formulated in such a way that its solution does not require a prior solution to the problem of induction. We can see the independence of the two problems if we cast the problem of success in the following form: why is it that many of the theories of the natural sciences enable us to predict nature and to intervene in the natural order in ways we want to so much more frequently and more accurately than (say) the theories of the ancient Greeks permitted them to do? This is clearly a *comparative* version of the problem of success. Its solution does not require us, as the problem of induction apparently does, to show that our theories are always (or even usually) reliable guides to the course of nature. The problem of success requires us only to explain why certain sorts of theories, authenticated by certain sorts of probative procedures, tend to promote certain cognitive ends more effectively than other sorts of theories, grounded in other forms of legitimation, do. As I shall be construing the problem of success in this section, it is fundamentally the challenge of explaining why certain modes of knowledge authentication produce more reliable results than others do.[26] We can thus leave the problem of induction in its general form conveniently to one side.

So let us now turn to my pair of stories. For the first, let us imagine that I am having problems getting my car started on a cold morning. My mechanic hauls it into the garage and replaces the brake drums, returning the car to me the next day. In the meantime, the weather takes a decided turn for the better. I crank up the engine, and the car starts without difficulty. When the mechanic bills me for replacing my brakes, and I complain that he did not do what he was supposed to, he replies by pointing out that my car starts now, and that that was what I wanted all along. Moreover, he points out that *all* the cars in his shop that day suffering from ignition problems had their brakes replaced and invariably the problem was solved. In sum, he claims that his tinkering with the brakes cured my starting problem, and cites as evidence for his claim the acknowledged fact that my car—along with the others suffering similar problems in his shop—now starts without difficulty. In exasperation, I explain to him that the state of the brake drums could have nothing whatever to do with the operation of the starter. He replies that such happens to be *my* theory about how cars work, but that *he* has a different theory, according to which brake wear can be a cause of poor ignition. Being a reasonable fellow of sorts, he cites as evidence for *his* theory the fact that the car started smoothly once the brakes were replaced. What am I to do in this case? Well, the first thing I might do is to point out that there is a different explanation than his for the sudden improvement in my starter's performance (namely, the warmer weather). Because there is, and because he hopes to get paid, he must show me some empirical evidence which supports his explanation of the starter's improvement rather than my meteorological hypothesis. If, moreover, I can point to plenty of other cars whose starting performance has improved dramatically with warming weather when no one was let loose on their brakes, my case is won. (At least as far as I and the Better Business Bureau are concerned.) Now, what is going on here? In effect, my mechanic and I are comparing *different probative strategies for the evaluation of beliefs.* My mechanic is evidently quite willing to shape his beliefs according to a simple *post hoc ergo propter hoc* policy. By contrast, I am insisting that discriminating tests be designed in order to rule out some of the many incompatible hypotheses which his strategy supports. (After all, my hypothesis is, like his, supported on *post hoc* grounds.) Beyond that, I can point to improved starting performance in other automobiles, whose brakes were not replaced. I deny that anyone who thinks carefully about these two strategies for the evaluation of empirical claims can have any doubts about which one is more likely to produce reliable results. My mechanic's failure to impose any form of experimental controls on his causal claims is likely to lead him to make far less reliable predictions than I will. In short, my strategy will save me from several sorts of failure to which my mechanic friend will sometimes fall prey. This is not to say that hypotheses which pass my sorts of tests will never be mistaken, nor that theories which pass his tests will never lead to correct predictions. It is simply to say that my strategy will produce conjectures which break down less frequently and less quickly than his will, and that is precisely why we say that one theory is more successful than another.

Consider, as a second example, the testing of a new drug said to be efficacious in curing arthritis. If I want to test its effectiveness, I might begin by enlisting a group of physicians who would prescribe the drug for their arthritic patients. Suppose, in the first run of the test, the only evidence reported back to me is that 55% of those who took the drug reported a reduced level of pain 24 hours after the onslaught of an acute attack of arthritis. Well, what conclusion will I draw? I might, if I am hasty, pronounce the drug a qualified success. But if I am the least bit careful, I will draw no such conclusion whatever, for the test itself is very badly designed. For all I know, for instance, it might well be that 55% (or more!) of patients who take no medication whatever also report improvement after 24 hours.

So, in the second stage of the testing, we need to devise a more complex experiment. We might divide the patients into two groups, administering no drugs to one group and the drug being tested to the other. Suppose, after the experiments are performed, that it emerges that 55% of those given the drug reported improvement, while only 20% of those given no treatment reported improvement. Well, these results are rather more impressive, but again, we have to be careful about drawing any conclusions about therapeutic efficacy. We have introduced certain controls on the experiment, it is true, but we can still imagine all sorts of ways in which the reported results might be compatible with the fact that the drug is of no therapeutic value at all. Specifically, given what we know about the placebo effect and the psychosomatic character of pain, it may well be that patients given any pill, even a worthless one, will report an improvement—just because they expect medication to make them better and that expectation itself will sometimes have the desired effect. Because the control group was given no pill at all, the different results in the two cases might have nothing whatever to do with the specific character of the drug under investigation.

Realizing this, we re-design our test. For the third run of it, we give pills to *both* the control group and the test group, but only the administering physicians know that the control group receives sugar pills. Suppose the results of this experiment are as follows: the group given the real drug reports a 55% improvement in 24 hours while the group given the placebos reports a 30% improvement. Well, the evidence for therapeutic value is getting more impressive, but there are still causes for concern about the significance of the results. We have learned from many studies that those conducting experiments often have a way of transmitting their knowledge and their expectations to the human subjects on whom they are experimenting. There are all sorts of documented forms of conscious and unconscious suggestion that might be going on, even though the doctors are conscientiously trying to treat the two groups of patients identically. In short, the physicians might be conveying to their patients their knowledge of which pills are placebos, or the doctors might

be interpreting their patients' comments so as to support their own expectations. If we want a strict test of the drug, we need to set up a situation where
those giving out the pills and interviewing the patients have no idea whether
the patients they are dealing with have been given the real drug or the placebo.
With suitable precautions, such an experiment can be devised; indeed, this
technique now represents a standard part of the repertoire for assessing the
efficacy of therapies of many sorts, whether drugs or psychoanalysis.

Those familiar with experimental design will recognize in the chronol-
ogy I have described the transition from an uncontrolled to a controlled experiment, from a controlled but unblinded experiment to a single-blind experiment, and finally from a single-blind to a double-blind experiment. For
each transition, we can lay out some good reasons to expect the results of the
later test to be more reliable than the results of the earlier one. Everyone who
studies experimental methods knows the story to be offered in each case. Thus,
in the first test (where we gave everyone the drug), we were using the simple
method of agreement. Once we introduced the control group, we were using
the joint method of agreement and difference. Anyone who thinks about these
two methods will realize why theories which stand up to the latter sort of
test are more likely to endure than theories which pass tests associated only
with the method of agreement. *Some* of the mistakes into which the unaided
method of agreement will lead us can be guarded against by using the two
methods in conjunction. Even those unfamiliar with scientific methods can
surely see why, if our concern is to find out whether the specific drug being
tested is efficacious, procedures such as control groups and blinding will allow
us better control over many extraneous variables and influences which can creep
into experimental design.

What such examples vividly illustrate is that we need not engage in "high
epistemology" to understand what is going on and why. The comparative
reliability of various testing procedures can be explained without resorting
to the realist's ambitious claims about the truthlikeness of scientific theories.[27]
We already have in hand an informal logic for testing causal hypotheses which
will rationalize and justify many of the methods of the natural sciences. (Every
good textbook on experimental design goes much further toward explaining
why science works than all the writings of scientific realists put together!)
So far as I can see, scientific realism is just not needed to give a viable account of those methods. The "logic" of theory testing, imperfect as it is, puts
us in a position to make some comparative judgments about the reliability
of various methods of inquiry and, via those judgments, we can explain why
theories which pass certain sorts of tests tend to endure longer than theories

which pass other, less demanding sorts of tests. The explanation of the success of science, I submit, is no more mysterious and no more elusive than that.

This explanation of the success of science has the added virtue of being straightforwardly testable. It predicts, for instance, that where there are individuals or whole societies which shape their beliefs without the controls associated with science, those beliefs will be less reliable on the whole than the beliefs of a "scientific" culture. To be more specific, it predicts for instance that the medical practices of so-called primitive societies will tend to be less reliable and less efficacious than the practices of modern Western medicine (provided, that is, that physicians in those societies use less robust testing procedures for their theories of disease than their Western counterparts do — which may or may not be so). This is not to assert that non-scientific cultures can never discover "cures" which have eluded Western medicine, since weak heuristic and probative methods are sometimes capable of producing useful discoveries. The claim here would rather be that the frequency and reliability of such "discoveries" should be lower in societies which do not use controlled methods than in societies which do. I do not have the evidence at hand to confirm such predictions. My only point in making them is to show that the explanation of success offered here is distinctly non-vacuous.

If there are those, like the relativists, who refuse to accept this account, they must, for instance, counter the claim that a controlled experiment is an improvement on an uncontrolled one, and they must establish that a double-blind procedure is no improvement over a single-blind one. They must make plausible their claim that such experimental techniques are nothing more than socially-sanctioned conventions whose limited validity, such as it is, applies only to our culture and our time. They must show that we are just kidding ourselves in thinking that we have learned something in the last 300 years about how to put questions to nature.

6. Conclusion

Science is successful, to the extent that it is successful, because scientific theories result from a winnowing process which is arguably more robust and more discriminating than other techniques we have found for checking our empirical conjectures about the physical world. On a case-by-case basis, we can usually indicate why these methods and procedures are more likely to produce reliable results than certain other methods are.[28] Those procedures are not guaranteed to produce true theories; indeed, they generally do *not* produce true theories. But they do tend to produce theories which are more reliable than theories selected by the other belief-forming policies we are aware of. The methods of science are not necessarily the best possible methods of inquiry (for how could we conceivably show that?), nor are the theories they pick out likely to be completely reliable. But we lose nothing by conceding that the methods of science are imperfect and that the theories of science are probably false. Even in this less-than-perfect state, we have an instrument of inquiry which is arguably a better device for picking out reliable theories than any other instrument we have yet devised for that purpose. We can explain in great detail why that instrument works better than its extant rivals. Because we can, the success of science ceases to be quite the mystery which some philosophers and sociologists have made it out to be.[29]

Notes

1. I am grateful to a variety of friends who commented on previous versions of this essay and helped clarify my thinking about several of its central themes. They include Alberto Coffa, Clark Glymour, Rick Creath, Tom Nickles, Peter Barker, Arthur Donovan, David Hull, Rachel Laudan, Adolf Grünbaum, Andrew Lugg, Robert Butts, Ilkka Niiniluoto, Nicholas Rescher, and Gary Gutting, as well as several of my immediate colleagues.

2. See letter xiv in Voltaire's *Lettres Philosophiques* (Rouen, 1734).

3. Careful readers will take exception to my juxtaposition of realism and relativism in this way, pointing out that they are really not so much "opposites", as they are orthogonal to one another. Strictly speaking, after all, those who deny realism tend to be instrumentalists or idealists (rather than relativists per se), while the natural opponents of relativism are what we might call "objectivists". Nonetheless, it is illuminating to play realist and relativist perspectives off against one another since (a) they are rival *epistemic* traditions which are genuine contraries of one another (i.e., they cannot both be correct), and (b) as I point out below, there is a widespread tendency to assume that weaknesses in either count as arguments for its rival.

4. To be more specific, the form of relativism which I shall be discussing chiefly involves the denial that any techniques or methods for warranting knowledge claims are "better" than any others. One might call this the thesis of 'methodological relativism'; it is different from, and much more ambitious than, the thesis (often called 'ontological relativism') to the effect that no ontological framework is "privileged".

5. See, for instance, B. Barnes and D. Bloor, "Relativism, Rationalism, and the Sociology of Knowledge", forthcoming.

6. As I shall show shortly, however, it is important not to draw too strong an analogy between judgments of rationality and judgments of success.

7. See, for example, Hilary Putnam's treatment of the success of science in his *Meaning and the Moral Sciences* (London, 1978).

8. I am grateful to Alberto Coffa (private correspondence) for persuading me of the urgency of avoiding pinning our characterizations of scientific success on the aims, real or avowed, of working scientists.

9. For a lengthy discussion of some of the changes which have taken place in the cognitive goals of scientists, see my *Science and Hypothesis* (Dordrecht, 1981) and *Science and Values* (Berkeley, forthcoming).

10. No claim about the relative success of science would be complete without reference to Paul Feyerabend's recent tirade against the thesis that science has been successful. In his *Science and a Free Society* (London, 1978), especially pp. 100ff., Feyerabend asserts that the results of science (i.e., its successes) are no more impressive than the successes achieved by the cosmologies of many "primitive" societies. Feyerabend goes on to claim that science *appears* to be more successful than other systems of nature only because of the systematic suppression of other approaches in our culture: *"Today science prevails not because of its comparative merits, but because the show has been rigged in its favour"* (p. 102; italics in original). Earlier ways of studying nature "have disappeared or deteriorated not because science was better but because *the apostles of science were the more determined conquerors . . .* [who] *materially suppressed* the bearers of alternative cultures" (p. 102; Feyerabend's italics). As he warms to his topic, his claims become even more vitriolic: "The superiority of science is the result not of research, or argument, it is the result of political, institutional, and even military pressures." *Ibid.* As with most of Feyerabend's more provocative theses, this is more bluster than substance. Pointing (quite rightly) to the fact that pre-scientific cultures have made many very useful discoveries about how to manipulate nature, he concludes that the ideologies of those cultures are as successful empirically as the theories of science (or that they would have been if we had not systematically eradicated their advocates). This, clearly, is a monumental *non sequitur.* The judgment that science is more successful (in the sense spelled out above) than the nature philosophies of other cultures is, as Feyerabend is more than clever enough to realize, entirely compatible with the claim that non- or pre-scientific cultures have produced theories and methods which sometimes work well for their purposes. But since an acknowledgment that science has been more successful than those rivals would undercut Feyerabend's epistemic anarchism, he conveniently fails to alert his readers to the fact that the slide from the claim that pre-scientific theories have enjoyed some successes to the thesis that those theories have been as successful as, or more successful than, science is a monumental piece of bad reasoning.

11. H. Putnam, *Mathematics, Matter and Method* (Cambridge, 1975), p. 73.

12. See references to, and criticism of, the work of these authors in my "A Confutation of Convergent Realism", *Philosophy of Science* 48 (1981): 19–49.

13. To say as much is probably to make life too easy for the realist. Since Duhem, it has been widely recognized that theories typically do not impinge on experience directly but only in conjunction with a wide variety of other auxiliary assumptions. Under such circumstances, it is entirely possible that a theory could be true *simpliciter,* and yet such that all the observed consequences we attribute to it (derived from conjunctions of that theory with auxiliary assumptions) could be false. (I owe this point to Jarrett Leplin.)

14. See my "A Confutation of Convergent Realism".

15. *Ibid.*

16. I shall resist the temptation to dwell on the fact that the relativist evidently *exempts* his own methods of theory evaluation from this general relativist critique. Since numerous authors have pointed to this apparently self-refuting feature of relativism, I shall not discuss it at length. (See, for instance, my "A Note on Collins's Blend of Relativism and Empiricism", *Social Studies of Science,* 12 (1982), 131–132.)

17. Perhaps the most strenuously relativist of this group is Harry Collins, who seem-

ingly denies that there is any sense in which we can say that science is successful in predicting the world. Because Collins believes that "the natural world has a small or non-existent role in the construction of scientific knowledge" (H. Collins, "Stages in the Empirical Program of Relativism", *Social Studies of Science* 11 (1981): 3) and that "reality [does nothing] to circumscribe possible individual beliefs" (H. Collins and G. Cox, "Recovering Relativity: Did Prophecy Fail?", *Social Studies of Science* 6 (1976): 437) he is presumably forced to deny that any meaning can be attached to the claim that any system of belief is any more successful than another.

18. In fact, I believe that the success of science does refute most of the extreme forms of relativism, but it would require an independent argument to show that, and its development would carry the narrative too far afield.

19. This is not to say that all parts of science are predictively reliable, nor even that science is always more reliable than other ways of second-guessing the future. The specific claim is a limited one: to wit, that certain scientific theories have a much better predictive track record than most of their extant, non-scientific counterparts.

20. In using this example, I do not mean to suggest that all the cognitive aims or interests associated with Western science correspond to identifiable technical interests which we can identify across a broad spectrum of cultures. I *do* mean to insist, however, that there are some interests which cut across boundaries of culture and society. It is an issue of great consequence, both intellectually and practically, whether the methods of warranting associated with science do or do not promote those aims.

21. Throughout this essay, I adopt three simplifying assumptions to make my task more manageable: (a) that there is a set of methods which we can identify as "scientific"; (b) that these methods are shared between proponents of rival theories; and (c) that these methods do not radically underdetermine theory choice. All these assumptions have been hotly contested by various relativists. I have elsewhere tried to defuse the force of the relativist's arguments on these issues. (I take on the first claim in "What Remains of the Scientific Method", forthcoming; the second in my *Science and Values* [Berkeley, forthcoming], and the third in my "Overestimating Underdetermination" forthcoming.) Readers will have to judge for themselves whether my arguments make it plausible for me to adopt here the simplifying assumptions indicated above.

22. I am not asserting categorically that Western hydrology would necessarily surpass all the rival, apparently "non-scientific" techniques for treating these phenomena. I do not know whether it would or not. I am, rather, showing how one might go about ascertaining which parts of science exhibit a degree of success which calls out for special explanation. (Obviously, should the folk wisdom of some societies produce theories which are consistently more successful than science, that would equally call for some form of special explanation.)

23. One sees precisely this assumption in the influential work of Mary Douglas. She argues, for instance, that "it is no more easy to defend . . . objective scientific truths than beliefs in gods and demons" (Mary Douglas, *Implicit Meanings* [London, 1975], p. xv). In effect, her analysis denies in principle that any methodological defense of the claims of science could be given which would show that those claims were better grounded than the beliefs of any non-scientific culture.

24. The relativist's cause is aided and abetted here from some unusual quarters, not least from the arch-rationalist, Imre Lakatos. He has claimed on numerous occasions that any point of view, however bizarre—if only it is provided with enough funds and talented advocates —can accumulate impressive empirical successes. (One should add, for the historical record,

that Lakatos never provided any evidence for this assertion of his. One suspects he threw it in as a sop to his relativist friend Paul Feyerabend!)

25. Popper wrote: "No theory of knowledge should attempt to explain why we are successful in our attempts to explain things" *Objective Knowledge* (Oxford, 1972). The problem, of course, is that if epistemology cannot illuminate that problem, it is not clear what interesting tasks would remain for epistemology.

26. When I say that one theory is more reliable than another, I simply mean to refer to the fact that one theory is apt to be more useful, to be able to digest a larger and more disparate range of phenomena before it breaks down, than a theory which is less reliable.

27. The astute reader may note that neither of my examples of testing procedures involved the testing of that sort of deep-structure theory which is beloved by realists. But that omission reflects no limitation on the explanatory technique sketched out here. Basically, we test our most deep-structure theories, and credit them with success or failure, in precisely the same way that we test theories which are "closer" to "observation" (such as the two examples I discussed). Thus, if one wanted to explain the relative success of the atomic theory, one might venture to show that the battery of tests to which that theory had been subjected was more demanding than the sorts of tests which (say) hermetical theories of chemical structure had passed.

28. My discussion of the last few pages probably suggests that the justification of these methods is more straightforward and less problematic than it actually is. It would be less than candid not to note that there is some serious disagreement about exactly what rationale to give for some of the standard procedures of empirical control. But I would claim that the broad outlines of a rationale for such methods are clearly understood; that we know what technical and justificational problems confront us, and that we have some ideas about how to resolve them. In no case does it seem that such justificatory moves require us to go in the direction of scientific realism, i.e., of basing our explanation of the success of the methods of science on the thesis that the theories which science produces are true or nearly true.

29. Before I close, it is worth noticing how the approach to the problem of success sketched here exhibits the gratuitousness of the realist's would-be solution to the problem. If we can explain why the methods of science are apt to produce theories which are more reliable than theories produced by other methods, then we need not commit ourselves, as the realist evidently must, to a dubious claim about the truth or truthlikeness of the theories of science. Going beyond that reliability to postulate that our theories correctly characterize the world via their deep-structural commitments is to assume both more, and less, than is necessary to explain why scientific theories work as well as they do.

Einstein's Realism*

Arthur Fine

"... the success of this mode of appraisal, in which one of the
criteria is model-fertility tested over the appropriate time-span, is
the best argument for taking the theoretical models of the scientist realistically. Not, of course, as description, but as metaphor,
which is the only way that complex things forever somewhat out
of reach *can* be known."

— Ernan McMullin (1976)

1. Introduction

Recent scholarship has begun to document Einstein's philosophical development away from the empiricist and positivist influences that marked his
early scientific thinking, and towards the realism that played an increasingly
central role in his conception of science, following the articulation of general
relativity.[1] In correspondence in 1948, Einstein reflects on his development
in a way that nicely confirms the scholarship. Referring to the period around
1905, and the origins of special relativity, he writes, "At that time my mode
of thinking was much nearer positivism than it was later on. . . . my departure from positivism came only when I worked out the general theory of relativity."[2] Recent discussion of Einstein's position on the quantum theory has
also begun to recognize the importance of his realism (and not just his interest
in a causal or determinist base) as a focal point of his attitude.[3] Here too Einstein's late correspondence supports the story. In 1950 he wrote, "In the center
of the problematic situation I see not so much the question of causality but
the question of reality (in a physical sense)."[4]

While there can be no doubt that Einstein turned away from positivism
to realism, or that realism was important in his thinking about the quantum
theory, there is considerable room for speculation concerning exactly what
Einstein's realism involves.[5] For, despite the many brief and casual references
to realism in his writings, Einstein was not disposed to discuss the various
elements of his realism in a systematic fashion. To understand Einstein's realism we must, I think, first construct it from his own scattered remarks. That
task of construction is the primary task of this essay.

106

2. Theorizing

To facilitate that task it will be useful to begin by focusing on some central features of Einstein's epistemological thinking, for these will help us organize his remarks on realism. There are two features that will prove especially useful for our purpose, Einstein's "entheorizing" and his holism. I coin the word 'entheorize' to describe the following move (one well-known in analytic philosophy): when asked whether such-and-so is the case one responds by shifting the question to asking instead whether a theory in which such-and-so is the case is a viable theory. This is a move entirely characteristic of Einstein's post-positivist thinking, and one that comes out clearly, for instance, in his treatment of causality. For example in 1952, commenting on a manuscript concerning the principle of sufficient reason, Einstein wrote to E. Zeisler recommending Hume's conclusion that we have no direct knowledge of causality. He says of causality, "We speak of it when we have accepted a theory in which connections are represented as rational. . . . For us causal connections only exist as features of the theoretical constructs."[6] Frequently, Einstein would cite Heisenberg's uncertainty formulas to ground the necessity for entheorizing the issue of causality. The following sequence of quotations shows some of the details of this path.

> Heisenberg has shown that the alternative as to whether causality (or determinism) holds, or fails to hold, is not *empirically determinable*.[7] On this account it can never be said with certainty whether the objective world "is causal". Instead one must ask whether a causal theory proves to be better than an acausal one.[8] Still it is really not clear when one should call a theory "causal". Instead of "causal" I would prefer to say: a theory whose fundamental laws make no use of probabilistic concepts.[9]

The upshot is to move the entire issue of causality out of the empirical realm, where it would be conceived of as more or less separately and directly subject to experimental test. Instead, one gets at the issue of causality by specifying what counts as a causal theory (namely, one with non-probabilistic laws), and one replaces questions about whether causality holds in nature by questions about which theory is better. These questions, as Einstein tells us in his "Autobiographical Notes", have two aspects, external verification and inner perfection (Schilpp, 1949, pp. 21–23). We need not further pursue the "weighing of incommensurables" (Schilpp, 1949, p. 23) that constitutes Einstein's ideas on theory choice. But we ought to notice that the device of entheorizing links up with the holism that, I think, is also crucial for understanding Einstein's brand of realism.

In the first instances, holism applies to the testing of hypotheses in a way that is usually associated with Duhem and Poincaré (and later with Quine). Here the idea is that the natural unit for experimental confrontation is an organic theory and not some separable hypothesis. This trend in Einstein's thinking is clear already in 1921, where his essay "Geometry and Experience" concedes to Poincaré that indeed only the total system of geometry-plus-physics is testable, at least "*sub specie aeterni*" (Einstein, 1921, p. 236). In his 1938 book with Infeld the idea is expressed this way: "It is really our whole system of guesses which is to be either proved or disproved by experiment. No one of the assumptions can be isolated for separate testing" (Einstein and Infeld, 1938, pp. 30–31).[10] When Einstein returns to the discussion of geometry in his "Reply to Criticisms" (Schilpp, 1949, pp. 676–679) he tries to turn the table on Poincaré's conventionalism by suggesting that not only testability but also the internal criterion of simplicity must apply to the *whole* system of physics-plus-geometry. And then Einstein goes on as the "non-positivist" to extend his holism from theory choice and simplicity to questions of meaning, for he suggests that the natural unit for issues relating to the meaning of individual concepts is also the whole theory. Perhaps this holistic idea has its roots in his earlier agreement with Schlick that the axioms of a system provide implicit definitions of its concepts (Einstein, 1921, p. 234), a doctrine that Einstein retains in his later discussion of the conventionality of the analytic/synthetic distinction (Einstein, 1936, p. 293). In any case, his "Replies" reiterate this holism over meanings in their curt dismissal of Bridgman's demand that each concept, separately, be given an operational definition (Schilpp, 1949, p. 679).[11] The sharp way that Einstein takes with operationalism may well reflect a harsh judgment that he makes on his earlier self. For instance in 1920 (or 1921) he wrote to Solovine espousing the orthodox operationalist line:

> Every physical concept must be given a definition such that one can in principle decide, in virtue of this definition, whether or not it applies in each particular case. (Solovine, 1956, p. 20)

By contrast in 1951, in the context of some of his reading of Maimonides, Einstein shows his holistic shift in attitude:

> The strangest thing in all this medieval literature is the conviction that if there is a word there must also be a clear meaning behind it, and the only problem is to find out that meaning.[12]

What emerged from this shift was a holism that entheorizes questions of meaning, along with questions of testability.

To see how this treatment of meanings operates, and to see just how serious Einstein is about it, I want to display and then discuss Einstein's explicit remarks about the concept of truth. For a realist, of course, we would expect truth to be understood as some sort of "correspondence with reality". But, as we shall see, Einstein's remarks suggest no such idea, and this will be an important sign that his realism may be quite different from the standard metaphysical view that bears that label.

There is, first of all, a set of four responses given by Einstein in 1929 and gathered together under the title "On Scientific Truth". Only the first response has to do with truth, and it is this:

> It is difficult even to attach a precise meaning to the term 'scientific truth'. Thus the meaning of the word 'truth' varies according to whether we deal with a fact of experience, a mathematical proposition, or a scientific theory. "Religious truth" conveys nothing clear to me at all. (Einstein, 1929, p. 261)

Twenty years later in his "Autobiographical Notes" Einstein interrupts the historical narrative to state what he calls his "epistemological credo". It contains the following paragraph relating to the concept of truth.

> A proposition is correct if, within a logical system, it is deduced according to the accepted logical rules. A system has truth-value [*Wahrheitsgehalt*] according to the certainty and completeness of its coordination-possibility to the totality of experience. A correct proposition borrows its "truth" from the truth-value of a system to which it belongs. (Schilpp, 1949, p. 13)

In 1951, Einstein wrote a brief letter (in English) on the same theme, and which I give here in full:

> Truth is a quality we attribute to propositions. When we attribute this label to a proposition we accept it for deduction. Deduction and generally the process of reasoning is our tool to bring cohesion to a world of perceptions. The label "true" is used in such a way that this purpose is served best.[13]

These remarks on truth defy the traditional philosophical categories of correspondence or coherence theories. Indeed, I don't think the remarks point to any genuine "theory" of truth at all. Rather they call our attention to structural features in the use of the concept, primarily the role of truth in logical inference. In the 1929 remarks, which appear to respond to questions about religion, Einstein catches hold of a certain indexical quality to truth judgments; i.e., the way those judgments depend on the larger context of inquiry. In the later remarks, this insight shifts in the direction of his entheorizing holism. That shift makes the truth of individual propositions subsidiary to their being an essential ingredient of a "true theory". But there is no special concept of a true theory apart from the ordinary scientific idea of a theory being observationally well-confirmed, or verified — ideally, by all possible observations ("its coordination-possibility to the totality of experience"). This idea is expressed by Einstein in a cryptic footnote to his article in the 1953 Born *Festschrift*, where he writes, "The linguistic connection between the concepts 'true' ['*wahr*'] and 'verified' ['*sich bewähren*'] is based on an intrinsic relationship" (Einstein, 1953, p. 34). I understand this as an expression of the idea that the truth of a proposition signifies nothing more than its role in an observationally verified system. In particular, there is nothing in Einstein's remarks to suggest the realist idea that the truth of a statement marks a special relationship that the statement bears to "reality", such as "correspondence" or "picturing", or the like.

I believe that Einstein's treatment of truth is entirely characteristic of his attitude towards the meaning of concepts in general. I would summarize it this way. When asked for the meaning of a concept (or statement) in science Einstein responds, first, by referring us to its role in systematic inquiry, and then (where appropriate), by pointing out the observational dividends that accrue to the theory that employs the concept and organizes the inquiry.

> . . . our concepts . . . and systems of concepts are human creations, self-sharpening tools whose warrant and value in the end rests on this, that they permit the coordination of experience "with dividends" ["*mit Vorteil*"]. . . . (Einstein, 1953, p. 34)

In general, no specific dividends are tied to any particular concept (or statement). Benefits, when they accrue, are credited to the theory as a whole. I think we might well adapt Russell's famous remark about mathematics to paraphrase Einstein's conception of science: according to Einstein, in science we can never know exactly what we mean, nor whether what we are saying is true.

3. Realism

We are now in a position to put together a picture of Einstein's realism. I shall begin right in the middle of one of his discussions of the quantum theory, for there we can observe him using the machinery laid out in the preceding section, and we can see very concretely how it shapes his realism.

The context is a preliminary skirmish in Einstein's battle to have the quantum theory seen as an incomplete description of individual systems.[14] In this particular discussion Einstein focuses attention on the decay of a radio-active atom. The uncertainty formulas for energy and time rule out the possibility of determining the exact time of a specific decay, since the decay is a change of energy state. The quantum theory only gives probabilities for what we will find if we look for the decay. Einstein points out, therefore, that *if* the atom does have a definite decay time, then the probabilistic quantum description must be counted as incomplete. But, one might well ask whether there actually *is* a definite time of decay. Such a thing is unobservable, but could it nevertheless be "real" (the scare quotes are Einstein's)? Einstein then continues:

> The justification of the constructs which represent "reality" for us, lies alone in their quality of making intelligible what is given by the senses. . . . Applied to the specifically chosen example this consideration tells us the following.
>
> One can not simply ask: "Does a definite moment for the decay of a single atom exist?" but rather instead "Within the framework of our total theoretical construction, is it reasonable to assume the existence of a definite moment for the decay of a single atom?" One cannot even ask what this supposition *means.* One can only ask whether or not such a supposition is reasonable in the context of the chosen conceptual system, with a view to its ability theoretically to grasp what is empirically given. (Schilpp, 1949, p. 669)

Here then we see how Einstein takes the concept of a "real" event, entheorizes it, and then refuses the question of its meaning (or significance), shifting instead to the question of the empirical adequacy of the relevant theory. He might have put it this way. We don't know what it means to say that the atom "really" has a definite decay time, nor do we know whether that is "true". We can only ask whether a theory that incorporates such decay times is a good theory, from an empirical point of view. If so, then we do say that the atom has a decay time, and we do count that as a description of reality (whatever that means). *These two moves, to entheorize concepts relating to "reality" and to refuse further inquiry into their significance (deflecting inquiry to the empirical adequacy of the whole theory) constitute the foundation of Einstein's realism.*

Before proceeding to fill in some more of the details of that realism I want to cite two further passages to illustrate these dual themes of entheorizing and meaning-avoidance. The first passage begins with a wonderful sentence that Einstein uses to summarize his debt to Kant. "The real is not given to us, but put to us (by way of a riddle) [*Das Wirkliche ist uns nicht gegeben, sondern aufgegeben (nach Art eines Rätsels)*]." Einstein comments:

This obviously means: There is a conceptual model [*Konstruktion*] for the comprehension of the inter-personal, whose authority lies solely in its verification
[*Bewährung*]. This conceptual model refers precisely to the "real" (by definition), and every further question concerning the "nature of the real" appears
empty. (Schilpp, 1949, p. 680)

The second passage is from his 1938 book with Infeld, and occurs just
after the remarks cited in section 2 about experiment confronting the theory
as a whole. The authors recite the tale, familiar to readers of Descartes, of
how building an explanatory theory is like trying to figure out the workings
of a closed watch:

If he is ingenious he may form some picture of a mechanism which could be
responsible for all the things he observes, but he may never be quite sure his
picture is the only one which could explain his observations. *He will never be
able to compare his picture with the real mechanism, and he cannot even imagine the
possibility or the meaning of such a comparison.* (Einstein and Infeld, 1938, p. 31)

The emphasis on this last sentence is mine, for I want to draw attention again
to the vehemence with which the idea of a "correspondence with reality" is
rejected, as both pointless and meaningless.

We have seen that Einstein's realism is arrived at by entheorizing "the
real". To say more about this realism, then, we must say something more
about what sort of theories are open to carry the idea of realism. That is,
I propose to treat the issue of realism the way Einstein treats the issue of causality. In that context Einstein entheorizes by specifying what counts as a causal
theory, and then agrees to let causality stand or fall according to whether causal
theories turn out to be better than non-causal ones. For realism, then, I will
try to begin by specifying what counts as a "realist theory". If we look at
Einstein's remarks on realism, I think the basic idea surfaces right away. Its
clearest and most succinct statement is this:

Physics is an attempt conceptually to grasp reality as it is thought independently of its being observed. In this sense one speaks of "physical reality". (Schilpp,
1949, p. 81)

The key realist idea here is that the conceptual model (or theory) is to
be understood as an attempt to treat things as we imagine they would be were
they not being observed. Of course such a theory must pass the test of empirical adequacy before one can be satisfied with it; that is, it must succeed
in encompassing the outcomes of all possible experiments (or observations).
Moreover one could imagine a theory that is empirically adequate, in this sense,

but not realist. This sort of theory seems to be what Einstein had in mind in 1922, in one of his earliest criticisms of Mach. "Mach's system studies the existing relations between data of experience; for Mach science is the totality of these relations. That point of view is wrong, and, in fact, what Mach has done is to make a catalogue, not a system."[15] Mach's catalogue leaves out the realist idea that one seeks to connect up the experimental data by treating them as data *about* an "observer-independent realm". It is interesting to note that Einstein's critique of Mach on this score antedates the development of the quantum theory. For the quantum theory, that developed in 1925 to 1927, was quickly interpreted in a "sterile positivist" (Einstein, 1953, p. 33) manner; i.e., as providing *no more than* a device for coordinating the outcomes of all conceivable experimental procedures. Some of Einstein's earliest concerns over the quantum theory were on just these realist grounds,[16] and his continuing criticism of what he graphically referred to as the quantum theorists' "epistemology-soaked orgy"[17] was that they were playing a "risky game . . . with reality".[18]

Thus Einstein seems to have had a reasonably clear picture of a particular sort of non-realist theory—some version of Mach's "catalogue"—by contrast with which we can get at this idea of a realist theory. Roughly speaking, a realist theory binds together the observable data by means of a particular sort of conceptual model; namely one whose basic concepts bear a standard interpretation that does not refer to observers, acts of observation, or the like. (I use the locution of 'standard interpretation' as a way of emphasizing what realism actually requires, but which no model can, by itself, guarantee. See my [1984a] for a discussion of the issue.) Despite the roughness of this characterization, I think it will do well enough for our purposes. We can test it, as Einstein's, by seeing that it makes a Machian "catalogue", which *only* coordinates the data (even with predictive success), non-realist. And this nonrealism extends to the quantum theory if we read the probabilistic formulas of that theory in the orthodox manner; that is, as merely probabilities for the results of observations (not for what is there independent of observations).

We can now proceed to entheorize various realist expressions. Thus "real" objects (e.g., events, properties, etc.) are objects described by the basic concepts of a realist theory. The "real external world" is itself just the structure posited by the conceptual model. What of "realism" as a doctrine? Here Einstein enters a caveat, for realism is not to stand or fall with any one realist theory. Rather, Einstein construes realism as a *program;* namely, as the program of trying to construct realist theories that, ideally, would be empirically adequate for all possible experimental data. He expresses this idea in his "Reply to Criticism" as follows, "After what has been said, the 'real' in physics is to be taken as a type of program, to which we are, however, not forced to cling *a priori*" (Schilpp, 1949, p. 674). In 1955 he reiterates this idea:

> It is basic for physics that one assumes a real world existing independently from any act of perception. But this we do not *know.* We take it only as a programme in our scientific endeavors. This programme is, of course, pre-scientific and our ordinary language is already based on it.[19]

It was clearly this conception of realism as a program that enabled Einstein to retain his realist orientation in the face of the mounting success of the non-realist quantum theory. Thus in 1949 Einstein's response to some questions raised by Adolf Grünbaum includes the remark, "In my opinion the positivistic tendency of physics to try to avoid the concept of reality is futile, even if it will take some years to realize this."[20] For Einstein, then, the success or failure of realism is a question of whether the program of doing physics by constructing realist theories is a progressive or a degenerative one.[21] But the progress (or not) of a program is a matter of integrating historical judgment over time and, logically speaking, the question cannot be settled by attending to the fate of the current theory in the field—no matter how successful (or unsuccessful) that theory is by itself.

It accords well with this construal of "realist theory", and of realism as a program for constructing such theories, that Einstein bases his continued adherence to realism on an appropriately historical argument. He treats the history of physics (from Newton to Maxwell to his own general relativity) as a triumph of the realist creed.[22] He refers to realism as "a program that was absolutely standard in the development of physical thought until the arrival of quantum theory" (Einstein, 1953, p. 34). Thus Einstein suggests that the program of realism, although not a conceptual necessity, has stood the test of time so far, and this grounds his belief that it will continue to do so. It is important to see that Einstein's realism is subject to the usual canons of scientific judgment, in the manner of any research program. Although it may well express a prejudice of Einstein's, his realism is not at all the silly metaphysics of "whether something one cannot know anything about exists all the same", a description supplied by Otto Stern and one that Pauli and Born were rather too eager to accept (Born, 1971, p. 223 and p. 227.)[23]

It is also important to see that Einstein counts the program of realism a successful one, at least up to the advent of the quantum theory, despite his own understanding (and emphasis) that the ontologies of the various theories within the realist program have made radical shifts over time. Roughly speaking, the "real objects" have changed from the point particles of Newton to the continuous fields of Maxwell-Lorentz, and later of Einstein. Such a radical shift of ontology, then, is quite compatible with Einstein's realism. For Einstein, realism does not involve any idea like that of the "successive approximation to reality". The conceptual objects of successive, successful realist theories can be as radically unlike one another as point particles are unlike continuous fields. There is, moreover, no entity (e.g., "the real, external world") that stands "outside" the theories and to which the sequence of conceptual objects could be compared (to see how well they "fit"). I should point out here that just as the "realist" idea of science making successively better approximations to reality is not part of Einstein's realism, neither is the pragmatist (especially Pierceian) idea of defining reality (and truth) by reference to the ideal limit in which continued inquiry would (supposedly) finally result. Einstein is properly skeptical about the idea of such an ideal limit. He holds that "our notions of physical reality can never be final" (Einstein, 1931, p. 266), that any system of concepts "will have validity only for a special field of application (i.e., there are no final categories in the sense of Kant)" (Einstein, 1936, p. 292). As to whether scientific theorizing "will ever result in a definitive system, if one is asked for his opinion, he is inclined to answer no"[24] (Einstein, 1936, p. 294). While there can be no doubt, as I shall discuss in section 7 below, that Einstein had faith in scientific progress and in the *motivational* power of pursuing the realist program, *as though it might have a limit,* these ideas are not built into Einstein's concept of realism, nor entailed by it.

4. Realist Theories

There are, however, other features of realism that *are* built into Einstein's concept. So far we have merely required that a realist theory be organized around a conceptual model of an observer-independent realm. I believe this is the core of Einstein's concept of a realist theory. But in most presentations of this concept Einstein links it with two others that must also be counted as central to his realist program. These are the ideas (1) that the conceptual model be a space/time representation and (2) that this representation be a causal one (i.e., one with strict = non-probabilistic laws). Here are some typical ways these ideas get linked:

> *1930.* Physics is the attempt at the conceptual construction of a model of the *real world,* as well as [*sowie*] its lawful structure.[25]

1940. Some physicists, among them myself, can not believe that we must abandon, actually and forever, the idea of direct representation of physical reality in space and time; or that we must accept the view that events in nature are analogous to a game of chance. (Einstein, 1940, p. 334)

1950. Summing up we may characterize the framework of physical thinking . . . as follows: There exists a physical reality independent of substantiation and perception. It can be completely comprehended by a theoretical construction which describes phenomena in space and time. . . . The laws of nature . . . imply complete causality. . . . Will this credo survive forever? It seems to me a smile is the best answer. (Einstein, 1950, p. 756 and p. 758).

The more complete version of the realist program, then, is to build theories in which "everything should lead back to conceptual objects in the realm of space and time and to lawlike relations that obtain for these objects" (Einstein, 1953, p. 34). Thus for Einstein the concept of a realist theory generally includes both the idea of a space/time representation and the idea of causality, along with that of observer-independence. These three components of realism are not logically connected. That is, a theory could incorporate any one (or any two) of them and not incorporate the other(s). In particular, one can have a theory that does not involve a space/time representation, but whose laws are non-probabilistic and whose basic concepts are understood as referring to observer-independent entities; i.e., a causal and observer-independent, but non-spatio-temporal theory. The difficulties presented by the quantum theory actually led Einstein to consider such a possibility.

In 1927 Einstein attempted to construct a space/time extension of Schrödinger's wave mechanics. In the unpublished draft for a presentation to the Prussian Academy (for May 5, 1927) he introduced his project as follows:

As is well-known, the opinion currently prevails that, in the sense of quantum mechanics, there does not exist a complete space/time description of the motion of a mechanical system. . . . Contrary to this, it will be shown in what follows that Schrödinger's wave mechanics suggests how every solution of the wave-equation corresponds unambiguously to the motion of a system.[26]

But Einstein abandoned this project, apparently because of difficulties in applying his scheme to coupled systems. In particular, Bothe seems to have called Einstein's attention to a consequence that, as Einstein puts it, must "from a physical standpoint" be rejected; namely, the scheme violated the requirement that the interpolated motion for a total system be a combination of the possible motions of its component subsystems. This difficulty over coupled systems may well have sparked the interest in the issue of "locality" in quantum mechanics that later surfaced in the famous EPR paper.[27] I think the failure of this space/time project did lead Einstein to take seriously the idea that the physics of the future may not be spatio-temporal at all.

In his review article of 1936, Einstein calls such a non-space/time physics "purely algebraical" and, because the mathematical concepts for such a theory had yet to be invented, in 1936 he rejects the idea as "an attempt to breathe in empty space" (Einstein, 1936, p. 319). Nearly twenty years later he is no more enthusiastic, and for exactly the same reason. "My opinion is that if the objective description through the field as an elementary concept is not possible, then one has to find a possibility to avoid the continuum (together with space and time) altogether. But I have not the slightest idea what kind of elementary concepts could be used in such a theory."[28] If we read these remarks in conjunction with his reply to Karl Menger in 1949 ("Adhering to the continuum originates with me not in a prejudice but arises out of the fact that I have been unable to think up anything organic to take its place" [Schilpp, 1949, p. 686]), then I think it clear that a non-spatio-temporal kind of realism (a "purely algebraical" realism) would be an acceptable alternative for Einstein to his own pet idea for a continuous field theory, even if one not so highly prized.

With regard to the acceptability of a realism that was not causal, however, I don't think that Einstein showed the same degree of tolerance. His commitment to causality comes out in one of his earliest public reactions to the quantum facts, the concluding passage of his 1927 essay on the occasion of the two hundreth anniversary of Newton's death, where he wrote:

> Many physicists maintain—and there are weighty arguments in their favor— that in the face of these facts not merely the differential law (i.e. "how the state of motion of a system gives rise to that which immediately follows it in time"] but the law of causation itself—hitherto the ultimate basic postulate of all natural science—has collapsed. Even the possibility of a spatio-temporal construction, which can be unambiguously coordinated with physical events, is denied. . . . Who would presume today to decide the question whether the law of causation and the differential law, these ultimate premises of the Newtonian view of nature, must definitely be abandoned? (Einstein, 1927, p. 261)

Notice here that although Einstein identifies both causality and a space/time representation as thrown into doubt by the quantum theory, it is only on causality that he focuses (for safeguarding, of course) in the final, rhetorical question. His letter to the British Royal Society, on the same occasion, also targets causality rather than the space/time representation, when Einstein closes his letter with, "May the spirit of Newton's method give us the power to restore unison between physical reality and the profoundest characteristic of Newton's teaching—strict causality."[29] Of course Einstein clearly recognized the possibility for an indeterministic physics, just as he recognized the possibility for a purely algebraic physics. But whereas he could (reluctantly) accept the idea of abandoning a space/time representation, he seems never to have reconciled himself to the idea of abandoning causality. In the early years of the quantum theory he expressed his recognition of the possibility of an indeterministic physics, and his rejection of it, this way:

> In itself it is already sufficiently interesting that a reasonable science can exist at all after dispensing with rigorous causality. It is furthermore not to be denied that this surrender has really led to important achievements in theoretical physics, but nevertheless I must confess that my scientific instinct reacts against foregoing the demand for strict causality. (Einstein, 1928, Column 4)

In his later correspondence the sticking point comes out as clear as could be. In 1945 he writes, "I don't believe that the fundamental physical laws may consist in relations between *probabilities* for the real things, but for relations concerning the things themselves."[30] What Einstein expresses here is an adherence to a causal realism, and he seems never to have seriously entertained a realism divorced from causality (in *his* sense of non-probabilistic laws).

In this regard we should be careful not to overemphasize the role of realism (i.e., observer-independence), as opposed to causality, in assessing Einstein's attitude toward the quantum theory. For if one examines the grounds that led Einstein to reject the acceptance of quantum mechanics *as a fundamental theory,* then these turn out to involve the statistical aspects of quantum theory every bit as much as the issue of observer-independence. That is, Einstein's concerns had *two* foci. The idea that realism *rather than* causality is at issue owes some currency, I think, to the publication of Pauli's remarks in the Born-Einstein correspondence, where Pauli says, "In particular, Einstein does not consider the concept of 'determinism' to be as fundamental as it is frequently held to be (as he told me emphatically many times). . . . Einstein's point of departure is 'realistic' rather than 'deterministic', which means that his philosophical prejudice is a different one" (Born, 1971, p. 221). Pauli then goes on to summarize one line of argument of Einstein's. The crucial point, and the one that Pauli does not accept, is Einstein's contention that

> a macro-body must *always* have a quasi-sharply-defined position in the "objective description of reality". . . . If one wants to assert that the description of a physical system by a ψ-function is *complete,* one has to rely on the fact that

in principle the natural laws only refer to the ensemble-description, which Einstein does not believe. . . . What *I* do not agree with is Einstein's reasoning [above] . . . (please note that the concept of 'determinism' does not occur in it at all!). (Pauli, in Born, 1971, p. 223)

If Pauli sees in this line of reasoning only realism and not determinism (i.e., 'causality', in Einstein's sense), that can only be because he has fallen into the trap that Einstein warned about in 1933, "If you want to find out anything from the theoretical physicists about the methods they use, I advise you to stick closely to one principle: don't listen to their words, fix your attention on their deeds" (Einstein, 1933, p. 270). No doubt Pauli correctly reports Einstein's insistence on realism as the central issue. But even in Pauli's reconstruction we can see that Einstein's line of thought involves causality as well. For the remark that Einstein does not believe that "in principle the natural laws only refer to the ensemble-description" is exactly an expression of Einstein's rejection of the idea that the probabilistic framework of quantum theory (the "ensemble-description") is to be accepted "in principle"; i.e., as a permanent part of fundamental physics. This is, of course, Einstein's adherence to causal (i.e., non-probabilistic) theories. The full idea that Pauli sketches in this passage is none other than Einstein's idea for natural laws (non-probabilistic) relating "real things", i.e., Einstein's idea of causal realism. Einstein's own expressions of the line of thought that Pauli tries to summarize are quite clear on this point.[31] In 1952 he put it to Besso this way:

> If one regards the method of the current quantum theory as in principle definitive, that means that one has to forego all claim to a complete description of real states of affairs. One can justify this renunciation if one accepts that there simply are no laws for real states of affairs, so that their complete description would be pointless. . . . Now, I can't reconcile myself to that. (Speziali, 1972, pp. 487–488)

Once again it is the conjunction of realism *and* causality ('laws' = non-probabilistic laws) that is characteristic of Einstein's thought. And, even in the same breath in which Einstein tells us that realism is more central than causality, he actually conjoins the two of them — almost as though he didn't notice that they are linked together.

> The question of "causality" is not actually central, rather the question of real existents [*realen Existierens*], and the question of whether there are some sort of strictly valid laws (not statistical) for a theoretically represented reality. (Letter to Besso, April 15, 1950; Speziali, 1971, p. 439)[32]

Similarly, in correspondence with Solovine where Einstein explains his entheorizing attitude towards *determinism,* he writes (on June 12, 1950):

> The question is whether or not the theoretical description of nature should be deterministic. Especially in that regard, in particular, there is the question of whether there is in general a conceptual picture of reality (for the single case) that is in principle complete and free of statistics. Only concerning this do opinions differ. (Solovine, 1956, p. 99)

I do not think that in these various passages Einstein is conflating causality and realism. Rather I take their constant conjunction in Einstein's remarks as a sign that *for him* the only sort of realism worth taking seriously, as a program for theory construction, was a realism built up by using strict, nonprobabilistic laws. Thus, unlike his attitude towards a space/time representation (which he clearly desired but could imagine doing science without) I believe that for Einstein causality was a *sine qua non* for a worthwhile program of realism. I should like to express this by saying that causality and observer-independence were *primary* features of Einstein's realism, whereas a space-time representation was an important but *secondary* feature.

5. Secondary Features

For a rounded picture of Einstein's realism I must mention two other secondary features. The first is connected, essentially, with a space/time representation. For Einstein makes it plain, in discussing his reservations about the quantum theory, that when real objects are represented in space and time they must satisfy a *principle of separation.*[33] In discussing the space/time framework, Einstein expresses the idea this way:

> It is . . . characteristic of these physical objects that they are thought of as arranged in a space-time continuum. An essential aspect of this arrangement of things . . . is that they lay claim, at a certain time, to an existence independent of one another, provided these objects "are situated in different parts of space". (Einstein, 1948; as translated in Born, 1971, p. 172)

In his "Autobiographical Notes", Einstein puts it like this:

> But on one supposition we should, in my opinion, absolutely hold fast: the real factual situation (or state) of the system S_1 is independent of what is done to the system S_2, which is spatially separated from the former. (Schilpp, 1949, p. 85)

Elsewhere I have discussed the role of this separation principle in Einstein's critique of the quantum theory (Fine, 1981) and the important distinction between Einstein's principle and the "locality" principles employed in the Bell no-hidden-variables literature (see Fine, 1984b, esp. Appendix). Here, I just want to point out how this principle depends on a prior commitment to a realist description (the "real factual situation [or state]") and to the causal nexus, which must *not* allow distant *real things* to have any immediate causal influences on one another. A realist space/time framework that would fail to satisfy the separation principle is one that Einstein ridicules as "telepathic" (Schilpp, 1949, p. 85) and "spooky" (Schilpp, 1949, p. 83).[34] It seems clear, then, that if we are to pursue realism in a space/time setting, Einstein requires that setting to respect his principle of separation. Separation is a necessary part of Einstein's conception of a space/time theory, although both the space/time representation and separation are secondary within the program of realism itself.

A further secondary feature of Einstein's realism is that, ideally, he would prefer the ontology of realist theories to be essentially monistic. That is, he considers it not really satisfactory for there to be more than one kind (or category) of real object; for instance, point-particles *and* continuous fields. His whole program for building a unified field theory is centered on the realization of this ideal.[35] His reductionist attitude towards psychology, and the mind-body problem, also seems to derive from this monism.[36] One might speculate that, perhaps, the deep source of this monistic attitude lies in Einstein's reading of (and "reverence" for) Spinoza.[37]

The "round" picture that we now have of Einstein's realism, then, is this. In the center are the linked, primary requirements of observer-independence and causality. Important, but not indispensable, are the secondary requirements of a spatio-temporal representation, which includes separation, and of monism. This whole circle of requirements, moreover, is not to be construed directly as a set of beliefs about nature, but rather it is to be entheorized; i.e., to be taken as a family of constraints on theories. Realism itself is to be understood as a program for constructing realist theories, so conceived. And realism is to be judged on exactly the same basis as any other program for theory construction in science. Whatever are the nuances and "incommensurable" elements that enter into such judgments, "in the end" they are understood by Einstein to be based on the instrumental success of the scientific endeavor — the "dividends" that it yields, in its "coordination of experience", in terms of novel, successful predictions. The particular tenets of realism that are built into a realist theory are themselves to be judged "true" just to the extent that we hold the theory to be well-confirmed observationally. But what the "truth" of these realist tenets amounts to ("concerning nature") is not a question that can be answered. It is, rather, to be deflected by turning instead to inquire about the instrumental success (or failure) of the program of realist theory construction that employs these tenets.

If this summary of Einstein's realism seems to trail off evasively, that is because, as I have emphasized, Einstein was deliberately evasive concerning the significance of his realism. And if this evasive side-stepping seems to make his realism a little paler and more shadowy than one might have expected, I'm afraid that too is because, in point of fact, Einstein's realism is not the robust metaphysical doctrine that one often associates with that label. What then are we to make of this frail creature, and is it a realism worthy of the name at all?

6. Was Einstein A Realist?

After this rather lengthy and detailed account of Einstein's realism, it may seem frivolous to be asking, at this point, whether Einstein was indeed a realist. Perhaps so, but then, as we have seen, Einstein's realism is a very peculiar kind of object, and one certainly ought to inquire how that object fits in with realism as it is more commonly understood. In the recent philosophical literature there are several positions that have been singled out as characteristically realist. One that Hilary Putnam has persuasively labeled "metaphysical realism" is centered on the belief in an external world, a world with a determinate observer-independent structure to which, increasingly, our scientific theories approximate by means of correspondence relations linking theories and the world.[38] I think this is close to the conception of realism that Gerald Holton ascribes to Einstein when he writes:

> In the end, Einstein came to embrace the view which many, and perhaps he himself, thought earlier he had eliminated from physics in his basic 1905 paper on relativity theory: that there exists an external objective physical reality which we may hope to grasp—not directly, empirically, or logically, or with fullest certainty, but at least by an intuitive leap, one that is guided by experience of the totality of sensible "facts". (Holton, 1973, p. 245)

There are, to be sure, many passages of Einstein's that express ideas close to that of such a metaphysical realism. But I think we ought to take Einstein seriously when he instructs us on how to read those passages; namely, that we are to entheorize the "realist" language and deflect questions of meaning (and "correspondence") into questions of empirical support for the theory as a whole. I have certainly emphasized these themes already, but perhaps one more quote would not be amiss. In a typical metaphysically realist passage, in an address at Columbia University, Einstein raises the question as to the purpose and meaning of science, rejects the ideal of a merely positivist catalogue, and then says:

> I do not believe, however, that so elementary an ideal could do much to kindle the investigator's passion from which really great achievements have arisen. Behind the tireless efforts of the investigator their lurks a stronger, more mysterious drive: it is existence and reality that one wishes to comprehend. (Einstein, 1934, p. 112)[39]

Here, indeed, we have Holton's Einstein, the metaphysical realist. The address continues, however:

> But one shrinks from the use of such words, for one soon gets into difficulties when one has to explain what is really meant by 'reality' and by 'comprehend' in such a general statement.

Thus Einstein deflects questions of meaning and, just as we would expect, goes on to entheorize and to focus on empirical adequacy instead:

> When we strip the statement of its mystical elements we mean that we are seeking for the simplest possible system of thought which will bind together the observed facts. (Einstein, 1934, p. 113)

These strategies of entheorizing and deflecting enable Einstein to use the vocabulary of metaphysical realism but to pull its metaphysical sting. For when we come to understand what Einstein means by his realist language it turns out to involve a program of theory construction that is to be judged on the same physical basis as any other scientific program, and hence not a "metaphysical" program at all. Similarly, I would argue, the sting of Einstein's "realism" is also pulled by these strategies. For realism is not a matter of words ('external world', 'real state of affairs', etc.) but rather a matter of what beliefs and commitments we have in uttering those words. Einstein is very clear in telling us that his commitments extend no more than to the pursuit of "realist" theories, and that his beliefs do not go beyond believing in the potential such theories have for organizing the observable data "with dividends". The metaphysical realist, however, has commitments and beliefs that go much further than these. Metaphysical realism involves precisely those elements that Einstein refers to (above) as "mystical", and that he is at great pains to strip off.

It is very tempting to call Einstein's realism a "nominal realism", for although Einstein uses a realist nomenclature he carefully instructs us not to understand it in the usual realist way. I shall point out below, however, that there is a feature of Einstein's realism that we have yet to come to terms with, and that will suggest a better label. To continue, then, exploring the "realism" of Einstein we might attend to another current candidate for what is characteristic of realism. I have in mind Bas van Fraassen's characterization of "scientific realism":

> Science aims to give us, in its theories, a literally true story of what the world is like; and acceptance of a scientific theory involves the belief that it is true. (van Fraassen, 1980, p. 8)

Since this account is already framed in terms of scientific theories, it holds out some immediate promise of touching Einstein's way of thinking. The key ideas here have to do with the *aims* of science (a literally true story about the world) and the nature of belief engendered by the *acceptance* of a theory (namely, belief that the theory is true). These concepts, having to do with the aims of science and the nature of theory-acceptance, also seem congenial to Einstein's conceptual framework. Moreover, in the passage just quoted from his Columbia address, for instance, Einstein does describe the aim of science as the comprehension of reality. This is much like the passage cited earlier from his "Autobiographical Notes", "Physics is an attempt conceptually to grasp reality . . ." (see section 3, above). On the surface, I don't think there is much distance between this conception of Einstein's and van Fraassen's formulation in terms of a "true story about what the world is like". However, if we push a little deeper, and inquire about what such a true story (or, rather, true theory) is for Einstein then, I think, we come to a parting of the ways. For, as I discussed in section 2, when Einstein attaches the concept of truth to a theory, what he means is just that what the theory says about all the *observable* features of the world will be confirmed ("the coordination of experience 'with dividends'"). But this explication of "true theory" is almost exactly what van Fraassen means when he says that a theory is "empirically adequate" (van Fraassen, 1980, p. 14). The difference between the literal truth of a theory and its empirical adequacy comes out if we attend to unobservable features described by a theory. Literal truth is just that (whatever "that" is). Empirical adequacy is something less. It allows us to accept statements about unobservables for logical processing and the like, but to hold as belief only that the theory as a whole will be confirmed by all possible observations. Einstein's way with the concept of truth is just the same. He would say, perhaps, that some statement involving unobservables "is true". But what he *means* is just that incorporating that statement in the relevant theory, and processing it logically there, leads to an empirically adequate story. (Recall his discussion of the decay time of an atom, in section 3.) Once again it is easy to be misled by Einstein's words, his insistence (e.g., concerning the decay time) that statements about unobservables can indeed be true. But if we follow his own directions to get at what he means, then the realist tone of his words gets damped by the empiricist constraints that he places on that meaning. Indeed it would not be too far off if we summarized Einstein's views this way: "Science aims to give us theories which are empirically adequate; and acceptance of a theory involves as belief only that it is empirically adequate."

Of course students of van Fraassen will recognize those words as his, and that passage in particular as van Fraassen's informal characterization of his own empiricist and non-realist position, which he calls "constructive empiricism" (van Fraassen, 1980, p. 12). My argument, then, is that if we understand Einstein in the way that he asks us to, his own realist-sounding language maps out a position closer to constructive empiricism than to either "metaphysical realism" or "scientific realism".

I think there is no backing away from the fact that Einstein's so-called "realism" has a deeply empiricist core that makes it a "realism" more nominal than real. This is, perhaps, not so surprising if we recall the intellectual debt to Hume and Mach that Einstein often acknowledged explicitly,[40] and that is implicit in much of his scientific writings. But then, too, we should not back away from the fact that Einstein did insist on using the nomenclature of realism, and that he explicitly opposed the positivism and empiricism that was mobilized, especially, in support of certain interpretations of the quantum theory.

His characteristic expressions of opposition were terms like "sterile positivism" or "senseless empiricism". We could understand the adjectives here ('sterile', 'senseless') as expressing Einstein's negative attitude towards the probable long-term success of an empiricist or positivist program for doing science. But we could also take his expression as applying to the feeling-tone involved in pursuing a positivist (or empiricist) program. It would, he might say, feel pointless to *do* "positive" science; there would be no underlying motivation or drive; it would not make science seem really worthwhile or meaningful.[41] I invent, on Einstein's behalf, but not much. For if we look again at his Columbia address then we see that whereas he quickly backs off the *cognitive* force of his realist remarks, he fully embraces the *motivational* force of that realism — especially by contrast with the program of positivism. I believe that here is the clue to Einstein's realist language, and that here we find what makes Einstein's realism more than nominal. To anticipate the answer, let me suggest the term 'motivational realism' for Einstein's view.

7. Motivational Realism

As early as 1918, Einstein's expressions of realism are presented in terms of motivations for the pursuit of science. In his moving and oft-quoted address in honor of Max Planck's sixtieth birthday, Einstein identifies both a "negative motive" (to "escape from everyday life") and a "positive" one (to "substitute" the (realist) "picture of the world" for "the world of experience, and thus to overcome it").[42] There too Einstein compares the realist attitude to a kind of "religious feeling" that drives the scientific effort with "no deliberate intention or program, but straight from the heart" (Einstein, 1918, p. 225 and p. 227). Einstein uses similar expressions in his "Autobiographical Notes" to describe the period of his childhood (before the age of twelve), when he was immersed in "the religious paradise of youth", as a "first attempt to free myself from the chains of the 'merely-personal'" (Schilpp, 1949, p. 5). His second attempt (around the age of twelve) was in fact his turning to science, when he began to feel that "out yonder there was this huge world, which exists independently of us human beings and which stands before us like a great, eternal riddle, at least partially accessible to our inspection and thinking" (Schilpp, 1949, p. 5). Note the realist language *and* the motivational context.

The idea of pursuing science as a replacement for the "religious paradise of youth" is not suggested, of course, in terms of cognitive appeal, but rather in terms of what motivates, enlivens and gives meaning to one's activities. In their late correspondence, Einstein's old friend Solovine chides him for using the word 'religious' in this context, to which Einstein replies (January 1, 1951):

> I have no better expression than the term 'religious' for this trust [*Vertrauen*] in the rational character of reality and in its being accessible, at least to some extent, to human reason. Where this feeling is absent, science degenerates into senseless (*geistlose*] empiricism. Too bad [*Es schert mich einen Teufel*] if the priests make capital out of it. Anyway, there is no cure for that [*ist kein Kraut gewachsen*]. (Solovine, 1956, p. 102)

In these various ways, from 1918 on, Einstein tells us that realism is the main motive that lies behind creative scientific work and makes it worth doing. But he clearly suggests that realism does not motivate scientific work in the manner of a deliberate intention or plan or duty (e.g., "to seek realist theories"). It is not as though we were first persuaded that realist theories are somehow best and then, on that account, rationally enjoined to pursue them. The operation of realism as a motive is, rather, like that of a "drive". It produces behavior that accords with and instantiates an inner "trust", both behavior and trust coming "straight from the heart". I think that in this way Einstein is placing realism among the pre-rational springs of human behavior (not, of course, among the irrational ones), those springs that we often conceive of not just as the source of creativity, but also as the source of deep satisfaction in creative endeavors.

These ways of characterizing the motivational aspects of realism suggest to me the ideas of depth psychology, the unconscious mechanisms that cause certain actions and lend them a certain feeling-tone. This association is, perhaps, not so removed from what would be acceptable to Einstein himself. In 1953 he wrote:

> Even if we think that much of Freud's theory may be mythology produced by a mighty imagination, nevertheless I believe that there seems to be much truth in the idea of suppression and inhibition, and in the fact that most of the causation in our mental life is not accessible to our consciousness.[43]

The particular psychoanalytic concept that seems to me most apt for Einstein's realism is the concept of an *image* — the complex ideal of the parent, rooted in the unconscious, elaborated by means of childhood fantasies, and bound with the strong affect of that childhood period. The parental imago lies behind and drives certain of our behavior just as Einstein tells us that realism "lurks behind" and drives our work as scientists. Properly integrated, the imago can be the source of deep commitment and of deep satisfaction, just like realism. The fantasy elements associated with the imago will, however, find some outward expression. In Einstein, I would suggest, the fantasy elements associated with his realist imago were expressed precisely by his realist vocabulary. Their fantasy nature, I believe, is marked clearly by Einstein's instant retreat from them just as soon as they are let out into the open. I think Einstein knew perfectly well that spinning the tales of "reality" was just letting childhood fantasies have their head. But he also knew that the affect bound to these fantasies was a factor in scientific life that must be given its due.

Einstein's realism comes out most clearly in his realist language. That language, I urge, must not be taken at face value. It does not mark a set of beliefs about "reality". It is, rather, the dues that Einstein felt worth paying for his passionate commitment to science, and for the meaning that scientific work gave to his life. Einstein's realism, then, is motivational. It is not aue-quately expressed by any special set of beliefs about the world, nor even by the injunction to pursue realist theories. Motivational realism is really not a doctrine but a way of being, the incorporation of a realist imago and its ex-pression in the activities of one's daily, scientific life. That this incorporation and way of life actually produces confirmed theories, and hence "knowledge" was — appropriately — considered by Einstein to be a "miracle", concerning which he wrote to Solovine (March 30, 1952):

> The curious thing is that we must be content with circumscribing the "miracle" without having any legitimate way to approach it. (Solovine, 1956, p. 115)

We should be wary, therefore, of any attempt to "legitimate" Einstein's realism by construing it as a set of realist doctrines (or beliefs). For Einstein, realism was motivational, and the language of realism was just his way of "cir-cumscribing the 'miracle'".

Notes

*Research for this essay was supported by the National Science Foundation, and by the Guggenheim Foundation. The paper was written during tenure of a Guggenheim Fellow-ship. I want to thank both foundations for their support. I am also grateful to Hebrew Uni-versity of Jerusalem for their kind permission to use material from the Einstein Archives. A copy of the archives is housed in the Seeley G. Mudd manuscript library at Princeton Univer-sity, and if I give no published source for a reference it can be found there. (Where I translate from archival material in German, I also give the German text; otherwise there is an English original for my citation.) My thanks to Dana Fine for his help with the archival MSS, to Don Moyer for copies of the *Nature* references (nn. 15 and 29), and to Micky Forbes for considerable "fine-tuning". I hope the essay itself will express sufficient "thanks" to various correspondents whose interest has also helped.
    1. See, e.g., "Mach, Einstein and the Search for Reality" in Holton (1973), pp. 219–259, and Barker (1981).
    2. Letters to D. S. Mackey, April 26 and May 28, 1948.
    3. See Pauli's sorting out of the misunderstanding between Einstein and Born (Born, 1971, pp. 221ff.), and see John Bell's comments on this (Bell, 1981, pp. 46–47). This is one of the themes in Brush (1980), and in Jammer (1982). My (1976), which is the (uncited) work that Jammer draws on, also focuses on Einstein's realism.
    4. Letter to J. Rothstein, May 22, 1950.
    5. See, e.g., Howard (1983).

6. Letter to E. Zeisler, December 10, 1952: *"Wir sprechen von ihm, wenn wir eine Theorie akzeptiert haben, in welcher sich der Zusammenhang als rationell darstellt. . . . Kausal-Zusammenhang für uns nur als Eigenschaft des theoretischen Constructs existiert."*

7. Letter to A. Lamouche, March 20, 1955: *"Heisenberg hat gezeigt, dass das Gelten oder Nicht-Gelten der Kausalität (bezw. Determinismus) Keine empirisch entscheidbar Alternativ ist."*

8. Letter to H. Titze, January 16, 1954: *"Ob die objektiv Welt 'kausal ist', kann deshalb nie mit Sicherheit gesagt werden. Wohl aber muss man fragen, ob eine kausale Theorie sich besser bewährt wie eine akausale."*

9. Letter to J. J. Fehr, March 25, 1952: *"Dabei ist es gar nicht klar, wann man eine Theorie 'kausal' nennen soll. Ich würde statt 'kausal' lieber sagen: eine Theorie deren Fundamentalgesetze nicht vom dem Wahrscheinlichkeitsbegriffe Gebrauch machen."*

10. In correspondence with Solovine concerning the French translation of this book Einstein points out that, apart from providing a needed source of income for Infeld, there was a special purpose to writing it; namely, to emphasize the role of a realist epistemology in the history of physics, as an antidote to the rising tide of positivism. Einstein also makes it plain that the epistemological aspects were worked on with special care (*"recht sorgfältig"*). See the letter to Solovine, April 24, 1938 (Solovine, 1956, p. 70). Pais (1982, p. 495) suggests that Einstein was not enthusiastic about the book, citing a remark from a letter to Infeld in 1941. ("One should not undertake anything which endangers the tenuous bridge of confidence between people.") But while that remark may throw some light on the inter-personal aspects of their collaboration, Einstein's only reservation that I know of concerning the actual content of the book is contained in a letter to Solovine on June 27, 1938, where he urges Solovine (in the French translation) to fix up a passage that refers misleadingly to the time at which the sun sets. Einstein also says that he prefers the German title (*Die Physik als Abenteuer der Erkenntnis* [*Physics as an Adventure in Knowledge*]) to the English, since it gives a certain psychological emphasis (Solovine, 1956, p. 72 and p. 74). I think we are, in fact, on solid ground in taking the epistemological point of view in this book as authentically Einstein's.

11. See Howard (1983) for further discussion of Einstein's holism.

12. Letter to Rabbi P. D. Bookstaber, August 24, 1951.

13. Letter to S. Candido, November 4, 1951.

14. By an "incomplete description" Einstein means that the probabilistic assertions provided by the theory do not exhaust all the relevant assertions about the actual, physical state of the system. See my (1981) for a discussion of Einstein's own exposition of this idea, and my (1984b) for a detailed treatment of Einstein's various lines of argument.

15. Reported in *Nature* 112 no. 2807 (August 18, 1923): 253.

16. See, for example, his letters to Sommerfeld of August 21 and of November 28 in 1926, in Hermann (1968). Almost certainly Einstein's realism here had to do with the possibility of a causal space/time representation, as I discuss in section 4. For some other early concerns of Einstein's see my (1976).

17. *"erkenntnistheorie — getränkte Orgie,"* in letter to Schrödinger on June 17, 1935.

18. Letter to Schrödinger, December 22, 1950, (Przibam, 1967, p. 39).

19. Letter to M. Laserna, January 8, 1955.

20. Letter to A. Grünbaum, December 2, 1949.

21. This is the terminology of Imré Lakatos's "methodology of scientific research programmes". I think that Lakatos's methodology provides a good framework for the issue of

"testability" raised by Einstein's realism. See Worrall and Currie (1978). Hacking (1979) is an excellent review of Lakatos's ideas.

22. See his essays on Newton (Einstein, 1927) and Maxwell (Einstein, 1931), and his "College of Surgeons" lecture (Einstein, 1950). As I pointed out in note 10, the triumph of realism is the main underlying theme of the book with Infeld (Einstein and Infeld, 1938).

23. Von Laue, in a letter to Einstein on August 8, 1934, quotes Otto Stern as saying, "You are just as reactionary as Einstein" ["*Sie sind ja noch reaktionärer als Einstein*"].

24. Howard (1983) gives further references for this attitude of Einstein's.

25. Letter to M. Schlick, November 28, 1930. Translated by Holton (1973), p. 243. (For '*sowie*' I have put 'as well as', where Holton used 'and'.)

26. This is from the fragment of a MS entitled "*Bestimmt Schrödingers Wellenmechanik die Bewegung eines Systems vollständig order nur im Sinne der Statistik?*" ["Does Schrödinger's wave mechanics determine the motion of a system completely, or only in the sense of statistics?"] in Kirsten and Treder (1979), p. 134. I want to thank Paul Forman for sending me a copy of the fragment. There is a typescript of the whole MS in the Archives (#2-100-1), which includes a postscript stating the difficulty over coupled systems, discovered by Bothe, that I point out below. At the very end Einstein alludes to a suggestion (by Grommer) for modifying the scheme to deal with this objection, and remarks that the modification should be tried out, first, on some examples. But Einstein never allowed this MS to be published, and so most likely Grommer's idea didn't work. In Born (1971), p. 96, the undated letter (#57) from Einstein refers to this MS. (It was written on the bottom of a letter from Ehrenfest that Einstein passed along to Born. Born does not give the date of Ehrenfest's letter, but it is the letter of April 16, 1927.) In a letter from Heisenberg to Einstein on May 19, 1927, Heisenberg says that he has heard about this idea of Einstein's from Born and Jordan, and then Heisenberg wonders whether it could be investigated experimentally. Einstein's remarks at the Solvay conference in October, 1927 contain no reference to this "*Bestimmt*" scheme. But they do suggest "locality" problems reminiscent of Bothe's objection. See my discussion of Einstein's remarks at that conference in my (1981), and my argument for why we must distinguish Einstein's ideas on locality from those current in the Bell literature, in my (1984b) — especially the Appendix. I should like to use this note on the "*Bestimmt*" MS to point out an error in my (1981) narrative. The MS I refer to there on p. 148 is this same 1927 one, and it does *not* contain the critique of quantum mechanics that I say it does. That critique, I'm afraid, was an artifact of my own rather garbled notes. Nonetheless the "worries" that I there attribute to Einstein (over completeness, locality and the classical limit) were certainly his, at that time as well as later, and they could even be read *between the lines* of "*Bestimmt*".

27. See my (1981) for a discussion of EPR, and Einstein's reservations about it. (See too above, note 26.)

28. Letter to D. Bohm, October 25, 1954. Quoted in Stachel (1983), who discusses and gives further references to Einstein's ideas about a "purely algebraical" physics. See Jammer (1982) on this point too.

29. A translation of this letter is given in Nature 119 (March 26, 1927): 467.

30. Letter to M. C. Coodall, September 10, 1945. This is virtually the same expression that occurs near the end of his Herbert Spencer lecture (Einstein, 1933), p. 276.

31. See Schilpp (1949), p. 672, and Born (1971), p. 188 and p. 209, and recall (see note 14 above) that Einstein's distinction between complete and incomplete descriptions involves precisely the question of whether statistics are fundamental (i.e., the question of causality).

32. The letter to Rothstein (see note 4, above) that also begins by emphasizing realism over causality proceeds, similarly, to bring the issue of causality right into the discussion.

33. Einstein's term is *"Trennungsprinzip"*. See my (1981).

34. In a letter to E. Cassirer on March 16, 1937, Einstein uses the phrase "a sort of 'telepathic' coupling" [*"eine Art 'telepathischer' Wechselwirkung"*] to show his rejection of interpreting quantum mechanics in a non-separable framework. This is the earliest use I have noticed of the particular language of telepathy, although the insistence on separability in quantum mechanics certainly goes back at least to the difficulty with his *"Bestimmt"* MS (see note 26, above). The letter of March 21, 1942, to C. Lanczos also uses the term "telepathic" in this context (Dukas and Hoffman, 1979, p. 68). See Earman (1984) for the varieties, and difficulties, of separability.

35. See Pais (1982), chapter 26, for a discussion of that program, and also Stachel (1983).

36. The opening passages of Einstein (1928) and the closing passages of Einstein (1950) give one a good feel for Einstein's monistic reductionism.

37. The connection with Spinoza's monism fairly jumps out from a 1937 "aphorism" of Einstein's recorded by Dukas and Hoffman (1979), p. 38. "Body and soul are not two different things, but only two different ways of perceiving the same thing. Similarly, physics and psychology are only different attempts to link our experience together by way of systematic thought." See Hoffman (1972), pp. 94–95, for Einstein's "reverence" for Spinoza, and his view of himself as a "disciple". The Spinoza connection, of course, involves determinism as well as monism.

38. This "metaphysical realism" is labelled and attacked in Putnam (1981). For a different line of attack see my (1984c), and for reservations about Putnam's positive view see my (1984d).

39. Stachel (1983) cites this as a lecture given at UCLA in 1932, and gives his own translation.

40. See, e.g., Schilpp (1949), p. 13 and p. 21ff. The letter to D. S. Mackey of April 26, 1948 (see note 2, above), is informative on this score, emphasizing especially the influence of Hume.

41. A similar feeling comes through in Einstein's castigation of the quantum theory in a letter to D. Lipkin on July 5, 1952: "This theory (the present quantum theory) reminds me a little of the system of delusions of an exceedingly intelligent paranoic, concocted of incoherent elements of thought."

42. Translated as "Principles of Research" (Einstein, 1918), the original title of this address was *"Motiv des Forschens"* or, loosely, "Motives for Research". Holton (1973), pp. 376–378, re-translates some passages from this talk, as does Howard (1983). I have used the "official" translation.

43. Letter to J. Pirone, November 6, 1953.

References

Barker, P. 1981. "Einstein's Later Philosophy", in P. Barker and C. G. Shugart, eds., *After Einstein*, pp. 133–145. Memphis: Memphis State University Press.

Bell, J. S. 1981. "Bertlmann's Socks and the Nature of Reality". *Journal de Physique* 42 Colloque C2, 41–61.

Born, M. 1971. *The Born-Einstein Letters.* New York: Walter and Co.

Brush, S. 1980. "The Chimerical Cat: Philosophy of Quantum Mechanics in Historical Perspective". *Social Studies of Science* 10: 393–447.

Dukas, H. and Hoffman, B. 1979. *Albert Einstein, The Human Side.* Princeton: Princeton University Press.

Earman, J. 1984. "What Is Locality?" Preprint.

Einstein, A. 1921. "Geometry and Experience", in Einstein (1954), pp. 232–246.

———. 1927. "The Mechanics of Newton and Their Influence on the Development of Theoretical Physics", in Einstein (1954), pp. 253–261.

———. 1928. "Fundamental Concepts of Physics and Their Most Recent Changes". *St. Louis Post-Dispatch,* Supplement for December 9, p. 7, columns 1–4.

———. 1929. "On Scientific Truth", in Einstein (1954), pp. 261–262.

———. 1931. "Maxwell's Influence on the Evolution of the Idea of Physical Reality", in Einstein (1954), pp. 266–270.

———. 1933. "On the Method of Theoretical Physics", in Einstein (1954), pp. 270–276.

———. 1934. *Essays in Science.* New York: Philosophical Library.

———. 1936. "Physics and Reality", in Einstein (1954), pp. 290–323.

———. 1940. "The Fundaments of Theoretical Physics", in Einstein (1954), pp. 323–335.

———. 1948. "Quanten-Mechanik und Wirklichkeit". *Dialectica* 2: 320–324. Translated ("Quantum Mechanics and Reality") in Born (1971), pp. 168–173.

———. 1950. "Physics, Philosophy and Scientific Progress". *Journal of the International College of Surgeons* XIV: 755–758.

———. 1953. "Elementare Überlegungen zur Interpretation der Grundlagen der Quanten-Mechanik", in *Scientific Papers Presented to Max Born,* pp. 33–40. Edinburgh: Oliver & Boyd.

———. 1954. *Ideas and Opinions.* New York: Crown Publishing Co.

Einstein, A. and Infeld, L. 1938. *The Evolution of Physics.* New York: Simon and Shuster.

Fine, A. 1976. "The Young Einstein and the Old Einstein", in R. B. Cohen *et al.,* eds., *Essays in Memory of Imre Lakatos,* pp. 107–121. Dordrecht: Reidel.

———. 1981. "Einstein's Critique of Quantum Theory", in P. Barker and C. G. Shugart, eds., *After Einstein,* pp. 147–159. Memphis: Memphis State University Press.

———. 1984a. "Is Scientific Realism Compatible with Quantum Physics?" *Nous,* forthcoming.

———. 1984b. "What is Einstein's Statistical Interpretation, or Is It Einstein for Whom Bell's Theorem Tolls?" *Topoi* 3.

———. 1984c. "The Natural Ontological Attitude", in J. Leplin, ed., *Scientific Realism.* Berkeley: University of California Press, forthcoming.

———. 1984d. "And Not Anti-Realism Either". *Nous* 18: 51–65.

Hacking, I. 1979. "Imre Lakatos's Philosophy of Science". *British Journal for the Philosophy of Science* 27: 329–362.

Hermann, A., ed. 1968. *Albert Einstein/Arnold Sommerfeld Briefwechsel.* Stuttgart: Schwabe & Co-Verlag.

Hoffman, B. 1972. *Albert Einstein, Creator and Rebel.* New York: Viking Press.

Holton, G. 1973. *Thematic Origins of Scientific Thought.* Cambridge, Mass.: Harvard University Press.

Howard, D. 1983. "Critique: What Kind of Realist Was Einstein?". Preprint.

Jammer, M. 1982. "Einstein and Quantum Physics", in G. Holton and Y. Elkana, eds., *Albert Einstein: Historical and Cultural Perspectives,* pp. 59–79. Princeton: Princeton University Press.

Kirsten, C. and Treder, H., eds. 1979. *Albert Einstein in Berlin, 1913–1933*. Volume 1. Berlin: Akademie-Verlag.

McMullin, E. 1976. "The Fertility of Theory and the Unit for Appraisal in Science", in R. S. Cohen *et al.*, eds., *Boston Studies in the Philosophy of Science, Vol. 39*, 395–432. Dordrecht: D. Reidel. Epigraph from p. 429.

Pais, A. 1982. *Subtle Is the Lord*. New York: Oxford University Press.

Przibam, K., ed. 1967. *Letters on Wave Mechanics*. New York: Philosophical Library.

Putnam, H. 1981. *Reason, Truth and History*. Cambridge: Cambridge University Press.

Schilpp, P. A., ed. 1949. *Albert Einstein: Philosopher Scientist*. LaSalle, Ill.: Open Court.

Solovine, M., ed. 1956. *Albert Einstein: Lettres a Maurice Solovine*. Paris: Gauthier-Villars.

Speziali, P., ed. 1972. *Albert Einstein/Michele Besso Correspondence 1903–1955*. Paris: Hermann.

Stachel, J. 1983. "Einstein and the Quantum", in R. Colodny, ed., *From Quarks to Quasars*. Pittsburgh: University of Pittsburgh Press.

van Fraassen, B. 1980. *The Scientific Image*. Oxford: Clarendon Press.

Worrall, J. and Currie, G., eds. 1978. *The Methodology of Scientific Research Programmes*. Cambridge: Cambridge University Press.

What Makes Physics' Objects Abstract?

NANCY CARTWRIGHT AND HENRY MENDELL

1. Introduction

The laser is a more concrete object than the damping reservoir, and the crystal diode rectifier than the rectifier. Why? Most obviously the laser is a fleshy object, rich in properties, whereas the damping reservoir is practically bare; the crystal diode rectifier is far more fully specified than the rectifier. Certainly sheer number of features matters in distinguishing the abstract from the concrete. But a far more significant difference depends on the kinds of features specified, and not on their number. Damping reservoirs, which we will discuss in detail in section 3, are objects identified entirely in terms of their output: they are supposed to eliminate correlations in systems they interact with. But nothing is specified about how they produce this result. Similarly, a rectifier is supposed to turn alternating currents into direct ones. The crystal diode rectifier does the same job, but in a highly specific way. We not only know more about the crystal diode rectifier than about the rectifier; what we know is how it works. We know its causal structure. This, we shall argue, is why the crystal diode rectifier is significantly more concrete than the rectifier, and the laser than the reservoir.

Ernan McMullin has stressed for a long time that causal structure, as opposed to empirical law, is the primary explanatory mechanism in theoretical physics. This is the key to understanding why causal structure makes an abstract object significantly more concrete. Aristotle taught that to understand the factors that provide the true scientific explanation of a phenomenon is to understand what that phenomenon really is. Even though we may no longer accept Aristotle's attempt to categorize explanatory features into four types, this is a doctrine that is still presumed in modern scientific inquiry. The connection with the abstract and the concrete is immediate. If the kinds of explanatory factors specified for one system are a subset of those for a second, the second is more concrete than the first. When more explanatory factors are provided by a description, the description specifies more fully what the system really is in the terms deemed relevant by the theory that supplies the explanation; and hence the object described becomes far more concrete. We shall argue then that objects without causal structures are abstract. Adding causal structure makes them concrete. This is because, following McMullin, the causal structure gives the true explanation for what the object does; and determining the true explanation fixes what the object is.

The fullest alternative account of the abstract and the concrete in science comes from Pierre Duhem, who used the distinction to attack scientific realism. Duhem drew a sharp line between the concepts of physics and the concepts that describe common facts. The concepts of common facts are concrete; physics' are abstract and symbolic. For Duhem, the abstractness of physics concepts arose from the mathematical requirements of a physics formulated in differential equations: these concepts are artificially precise, in contrast to the concepts of common facts, which can be as imprecise and fuzzy-boundaried as the facts themselves. Duhem's claim that the concepts of physics are abstract served his anti-realism. The laws of physics, says Duhem, are neither true nor false. They are mere symbolic formulae. This is in part because the concepts, being abstract, do not describe real things; they describe only imaginary constructions.

But abstractness and scientific realism are different issues, and not all varieties of abstractness bear equally on questions of descriptive completeness, accuracy, and truth. This is so with our notion, where objects become more and more abstract as less and less explanatory information about them is given. Duhem was engaged in a debate about the roles of differential equations, boundary conditions, and models. He worried about how physics, with its mathematical constraints, could represent. We are concerned, not with representation but with explanation; and, we shall argue, our concept of abstractness and Duhem's are in perpendicular dimensions. Duhem scales abstractness according to descriptive completeness; we according to explanatory completeness. The two, we will claim, may vary independently.

2. Alternative Ideas of the Abstract

Pierre Duhem

Why do we describe some systems as abstract and others as concrete in a scientific theory? Pierre Duhem offers one answer. The concepts of common facts are concrete; those of physics are abstract and symbolic:

> When a physicist does an experiment two very distinct kinds of representations of the instruments on which he is working fill his mind: one is the image of the concrete instrument that he manipulates in reality; the other is a schematic model of the same instrument, constructed with the aid of symbols supplied by theories; and it is on this ideal and symbolic instrument that he does his reasoning, and it is to it that he applies the laws and formulas of physics.[1]

The two instruments can never collapse into one:

> Between an abstract symbol and a concrete fact there may be a correspondence, but there cannot be complete parity; the abstract symbol cannot be the adequate representation of the concrete fact, and the concrete fact cannot be the exact representation of the abstract symbol.[2]

For Duhem, a single theoretical fact may correspond to a variety of concrete facts. For example, the theoretical fact, "The current is on," may correspond to a collection of concrete facts: the displacement of a spot in a galvanometer, the bubbling in a gas voltameter, the glow of an incandescent lamp inserted in the wire, and so forth. Conversely:

> The same group of concrete facts may be made to correspond in general not with a single symbolic judgment but with an infinity of judgments different from one another and logically in contradiction with one another.[3]

In current philosophical jargon, the concrete facts underdetermine the theoretical. The simplest case has to do with the precision of measuring instruments: *any* of the incompatible values for a theoretical quantity that lie within the range "of error" of the most precise instrument used correspond equally well to the facts available.

The abstractness of the objects about which physicists reason in contrast to the objects they manipulate comes from two sources. One is connected with issues of wholism, which we commonly associate with Duhem: theoretical concepts are abstract because they can only be applied to reality by presupposing a number of theoretical laws and rules of correction. The second source is connected with the methods modern physics uses to represent reality. Physics aims at simplicity of representation. But nature as it comes is complex and intricate. Hence there inevitably arises a mismatch between the abstract-theoretical representation and the concrete situations represented. The result is that the abstract formulae do not describe reality but imaginary constructions. They are best judged neither true nor false of reality itself.

In a paper we shall discuss more thoroughly in section 4, Ernan Mc-Mullin describes this mismatch: "The Galilean model of idealization to which the material sciences have been committed since the 17th century implies an initial separation of theory from the causally complex 'material' situation."[4] Duhem differs from the realist over the term 'initial'. For Duhem the theoretical aims of unification and simplification will always and unavoidably imply a separation of theory for the causally complex "material" situation.

An even more important problem than simplicity of representation for Duhem is the mathematical character of physics. The concepts of physics are precise, but reality itself does not exhibit this precise, quantitative nature:

> Let us put ourselves in front of a real concrete gas to which we wish to apply Mariotte's (Boyle's) law; we shall not be dealing with a certain concrete temperature, but with some more or less warm gas; we shall not be facing a certain particular pressure embodying the general idea of pressure, but a certain pump on which a weight is brought to bear in a certain manner. No doubt, a certain temperature corresponds to this more or less warm gas, and a certain pressure corresponds to this effort exerted on the pump, but this correspondence is that of a sign to the thing signified and replaced by it, or of a reality to the symbol representing it.[5]

Notice here that Duhem is talking about the theoretical concepts of a purely macroscopic physics, and is not concerning himself with the status of tiny, unobservable entities. A second example that Duhem uses is the sun. In order to find a "precise law of the motion of the sun seen from Paris":[6]

> The real sun, despite the irregularities of its surface, the enormous protuberances it has, will be replaced by a geometrically perfect sphere, and it is the position of the center of this ideal sphere that these theories will try to determine. . . .[7]

This kind of falsification is for Duhem an unavoidable consequence of the mathematical formulation of physics. He seems here to be echoing the earlier criticisms of Laplace and Poisson against Fourier's elevation of differential equations to first place in physics. Differential equations require sharp geometric boundaries.[8] So long as the differential equations are taken, as by Laplace, merely as convenient summaries of results that can be vigorously established by applying integral equations to the interaction of molecules, the reality of these boundary conditions is irrelevant. But the issue can no longer be ignored if we drop the molecular basis and adopt the differential equations as primary. When we do so, we inevitably end up with a physics whose concepts no longer describe reality, but are irreducibly "abstract and symbolic".

Duhem describes one important notion of abstractness, a notion which serves to separate theoretical from practical facts. But that cannot be all there is to the concept of abstraction in the physical sciences. For theoretical objects differ among themselves; some are more abstract than others. The concept of a damping reservoir, for example, intuitively seems considerably more abstract than that of a laser. This is an example we will discuss in detail in section 3. Here we remark only that Duhem's distinction does not separate the two. Both reservoirs and lasers fall equally on the same side of Duhem's divide. Both a laser and a reservoir may be viewed as "a concrete instrument that he [the experimental physicist] manipulates in reality". On the other hand, both may be seen as a "symbolic and ideal instrument" to which the experimentalist "applies the laws and formulae of physics". From either point of view, lasers and reservoirs come out exactly the same.

This should not be surprising since the sources of abstractness beset the theoretical treatments of lasers and reservoirs equally. Both are highly "theory-laden" concepts. They make sense as descriptions of the world only if a large number of theoretical laws that we believe to be true are true. Secondly, the laser and the reservoir to which the laws of physics are applied, as opposed to lasers and reservoirs in the laboratory, are both highly simplified and idealized objects. The example of lasers and reservoirs is discussed in Cartwright's *How the Laws of Physics Lie;*[9] the reader can look there to see some of the kinds of idealizations involved. In particular, the theoretical treatments of both systems involve the kind of artificial precision and geometrization that we have seen to be required by theories formulated in mathematical terms, for example, by the use of differential equations. Perhaps, somehow, *more* simplifications or more mathematical artifices are used in the theoretical description of reservoirs than in lasers. But it is notoriously difficult to devise ways of counting such things, and it seems, moreover, that there is a qualitative, and *not just* a quantitative difference between the two kinds of systems. It is this qualitative difference which we shall try to capture. For this we shall have to look elsewhere, to alternative concepts of abstractness.

Two Classical Notions

Classically, there are two important and opposing notions of the abstract, one of which we may associate with Platonism and the other with those schools which attempt to preserve Platonic discourse without Platonic entities, and in particular with Aristotle.

Perhaps the most common view of abstract entities has been that they are single properties distinct from all physical and concrete objects in which they are instantiated.[10] In accordance with this view of abstract objects, there arises the notion of a geometric space separate from the concrete, which we find attributed to Plato and to Plato's associate, Speusippus.

What is important about separate abstractions is that they do not have to be real properties of concrete objects, although they are normally conceived as related somehow to concrete objects. A Platonist, for example, will say that the concrete object is a defective sample of the separate abstraction. Abstract objects as separate entities are well suited to Platonism or to a conceptualism such as Locke's, where abstract objects are subjective. Presumably most modern attacks on abstract objects are not attacks on abstract objects *per se,* but on abstract objects as separated.

Duhem's abstract models are imaginary constructions which do not describe reality but are somehow taken from reality. They are modern-day separated abstract objects. Hence, generalizing the argument of the previous section, it is easy to see that the notion of abstract objects as separated is ill-suited to distinguish lasers from reservoirs. The difficulty is again that reservoirs are no more abstract in this sense than lasers, and lasers are no more concrete than reservoirs.

Aristotle raised many of the familiar problems about separated abstract objects: their independent existence, their relationship to the world as perceived, and even the problems of how a geometric space could be true of the physical world. His own, the original notion of abstract objects, avoided these problems. Is Aristotle's the notion of abstraction we want?

Scattered throughout the works of Aristotle we find expressions such as "things in abstraction" or "things from abstraction".[11] Here 'abstraction' means "taking away" or "subtraction". According to Aristotle many sciences study objects left from subtracted properties. We begin with a particular substance and all its properties, both accidental and essential. If we are examining the object as a triangle in the context of a geometrical discussion, we begin, say, with a triangular marking on a slate. We subtract the stone and chalk, the color, and other properties incidental to being a triangle. What is left when this psychological process is complete is the triangle treated as if it were substance. The subtracted properties may be conceived as incidental to what the object is treated as, namely a triangle. Of course what the object actually is, a stone slab, will be among the accidents of the triangle. In other words abstraction amounts to a psychological rearrangement of the structure of the slate.

The Aristotelian notion of abstraction has a distinct advantage over those treatments of abstract entities already discussed, for it provides an anti-Platonist means of identifying the object of a mathematical science such as physics. In its Berkeleyan guise it serves as a foil to Locke. It seems to provide a basis for a philosophy of science which can answer the question, "How does science pertain to reality?" and which eschews Platonic entities. For the properties examined in abstraction are just properties of actual substances.[12]

Unfortunately, the Aristotelian idea doesn't work to make the connection between laboratory lasers on the one hand and the laser which physicists reason about on the other. For it ignores the problems introduced by mathematical representation and the Galilean model of idealization that McMullin reminds us is characteristic of physics after the scientific revolution. Scientific theories do not just isolate objects of study; they also make false assumptions about them. They subtract not only incidental properties, but inconvenient ones as well, and replace them with others more amenable to calculation or which we already know how to treat in our theories. In fact, Aristotle's science did the same. Idealizations had no place in his philosophy; but they did in his practice.[13]

Even if we could solve the problem of idealization, we would seem to be involved in the awkward question of counting if we used Aristotle's concept of abstraction to judge lasers and reservoirs. For Aristotle, the less abstract object is closer to substance than the more abstract. How are lasers closer to substance than reservoirs? By virtue of having fewer unessential properties stripped away? If we adopted this view, we would be faced with the unpleasant task of figuring out how to count properties. A simple example illustrates the difficulties. Suppose we examine two markings on the blackboard, one triangular, the other quadrilateral. We examine the first as a right triangle, the second as a rhombus. To examine the right triangle we must subtract lengths of the sides, and the two acute angles, except such properties as are necessary for the markings to be a right triangle. To examine the rhombus, we subtract lengths of the sides, except the relative lengths of opposite sides, and angles, except the relative size of opposite angles. Are rhombi more or less abstract than right triangles? The question, as would any answer to it, seems pedantic and baroque. For reasons such as this it is not possible to develop a coherent total ordering for Aristotelian abstract objects. Some entities, such as rhombi and right triangles, are not naturally ordered with respect to each other.

But there are natural ways of partially ordering the abstract. In his discussions of different sciences whose objects are known in abstraction Aristotle suggests a simple criterion for ordering the relatively abstract and concrete. For example, the objects of geometry and mathematical astronomy are both known in abstraction. But the latter are nearer to substance. Why? It is not because if we count the properties subtracted in each case we see one set is smaller than the other. Rather, since geometry is used in mathematical astronomy, in doing mathematical astronomy we do not subtract geometrical properties. But in doing geometry we subtract some properties which are in mathematical astronomy.

For example, mathematical astronomy treats of the movement of stars. A geometrical treatment of the same objects would not consider the movement, though it might treat the stars as fixed points, and the paths of the stars as circular lines. Hence, Aristotle claims, mathematical astronomy is nearer to substance. For it works with a more complete description of an object, and so comes closer to a description of what the object is.

Even within a science it is possible to see that certain objects are more abstract than others. Triangles are more abstract than right triangles, since the right angle of the right triangle is subtracted when it is treated just as a triangle. Yet all the properties of triangles apply to right triangles.

We have, therefore, a means for partially ordering entities: A is a more abstract object than B if the essential properties, those in the description of A, are a proper subset of the essential properties of B.

In a similar way Aristotle orders sciences. Mathematical astronomy studies geometrical properties of certain moving points. We can distinguish those properties which pertain to movement and those which pertain to geometrical spaces. These constitute two different classes which are basic to human understanding. Thereby we can claim that the objects of science A are more abstract than those of science B if the set of kinds of properties treated by A is a proper subset of the kinds of properties treated by B. The basic insight is that where there are two clear classes of properties, the entity with properties from one class will be more abstract than the entity with properties from both. Within abstraction by subtraction, we can get a partial ordering of entities with reference to nesting of their kinds of properties.

The idea that we can partially order the abstract with reference to the nesting of kinds of properties is intuitively reasonable since the entity described by more kinds of properties, in this very straightforward way, does seem to be more completely described. We cannot, however, literally apply these two Aristotelian orderings to the case of reservoirs and lasers: their properties are not nested, nor are they items in different sciences in a hierarchical ordering.

Neither the Platonic nor the Aristotelian notion of abstractness seems to provide a means for explaining why lasers are concrete and reservoirs are abstract. Abstraction as separation did not work, because lasers and reservoirs fell in the same ontological category, whether abstract or concrete. Abstraction as subtraction failed, because lasers and reservoirs did not seem to be described in ways which would make them comparably abstract. On the other hand, the Aristotelian notion of abstraction was accompanied by a method of comparing the abstractness of different objects, namely via the nesting of properties, and, more important, via the nesting of kinds of properties. In section 5 we will exploit this basic intuition to develop a new notion of abstractness which fits the contrast in our paradigmatic case between lasers and reservoirs. But first we will isolate in the next section what we think the significant difference is that we are trying to illustrate with our examples of lasers and reservoirs.

3. Lasers and Reservoirs

There is one obvious way in which the objects that physicists reason about may be abstract. They may lack stuff, and be made of no material. Contrast helium atoms, infrared radiation, and electrons on the one hand, with levers, oscillators, and reservoirs on the other. The first are characterized in large part by what kinds of stuff they are made from; two electrons, two protons and two neutrons; or electromagnetic waves between .75 and 1000 microns in length; or negatively charged matter. But its stuff is no part of what makes an oscillator an oscillator or a lever, a lever. These objects are characterized in terms of what they do, and they may be made of any material that allows them to do what they are supposed to.

Reservoirs are another good example of this second kind. Reservoirs play a role in a variety of physics theories, especially thermodynamics and electromagnetics. Fundamentally reservoirs are large objects, in one way or another, that swamp the systems they interact with. With respect to electricity, the earth is a reservoir that "grounds" moving charge; in thermodynamics, a heat reservoir will give up heat until it finally brings the system it interacts with to its own temperature.

We will concentrate on reservoirs in quantum optics. There a reservoir serves to erase the memory of the systems it interacts with. The reservoir is given only enough formal characteristics to justify this outcome mathematically. To erase the memory in a system, one wants to introduce a Markov approximation there, and the interaction with the reservoir licenses this. The reservoir is assumed to have a very great number of degrees of freedom. Because it is large and chaotic, correlations among its own states persist for only a very short time, and it in turn destroys correlations in the systems it couples to. Nothing else is necessary for an object to be a reservoir.

The reservoir is a paradigm "black box". We know only its final effects, but nothing of what exists inside. Intuitively, it is a very abstract object. But what sense of abstractness expresses this intuition? Reservoirs undoubtedly lack stuff. Sometimes the electromagnetic field may serve as a reservoir, for oscillating atoms for instance; sometimes the walls of a cavity; sometimes the environment in a room. But a large number of objects that physicists reason about lack stuff, and yet most are far more concrete than the reservoir. Consider a laser. Lasers are supposed to amplify light and produce a highly coherent beam of radiation. They too, like the reservoir, can come in a variety of substances. They may be made of rare earth ions floating in dissolved molecules, or from organic dye solutions, or from ruby rods, or numerous other materials. Yet certainly there is a sense in which the laser is a far more concrete object than a reservoir. It is that sense of *concrete* that we are trying to explain.

What is it that makes a laser a laser? The answer to this question will provide the key to our problem about the abstract and the concrete. The first operating laser was a ruby laser made in 1960 by T. H. Maimon. Any introductory discussion will point out that ruby is not essential. For instance:

> Since 1960 many different kinds of lasers have been made. They do not all contain ruby, nor do they all use a flashtube [as Maimon's did].
>
> But what they do have in common is an *active material* (e.g., the ruby) to convert some of the energy into laser light; a *pumping source* (e.g., the flashtube) to provide energy; and mirrors, one of which is semi-transparent, to make the beam traverse the active material many times and so become greatly amplified.[14]

We see that a laser consists of at least three distinct components: the active material, the pump, and the mirrors. A more theoretical treatment will also add the damping reservoir as another essential component in this list. In one sense then the specific make-up of the laser is irrelevant: no specific material is required. But in another sense, the make-up is essential. The components are a part of the characterization of a laser as a laser.

But the specification of the components is not sufficient. Lasers work in a very special way. In any normal circumstances, in a population of atoms far more are in the ground state than in the excited state. The trick in making a laser is to create a "population inversion", in which the situation is reversed. Quantum theory then teaches that the radiation from an initial atom de-exciting will stimulate others to radiate as well, and the radiation that is produced in this way will be highly coherent, unlike light from ordinary sources. This method of production is even referred to in the name: 'laser' stands for "*l*ight *a*mplification by *s*timulated *e*mission of *r*adiation".

It is, then, part of the concept of the laser that it consists of a specific set of components which produce highly coherent light by a specific causal process. The method of operation *for* a reservoir, by contrast, is completely unspecified. The relevant degrees of freedom that are represented mathematically are not identified with any concrete quantities, and no account is given of the process by which these quantities couple to the littler system and destroy its correlations. This, we think, is why a reservoir is intuitively so much more abstract than a laser. Reservoirs are specified only in terms of their outputs, whereas lasers are specified not just in terms of their outputs, but also in terms of the constituent parts and the causal processes by which these outcomes are produced.

As hypothesized, this variety of abstractness is indeed independent of the one that concerns Duhem. For Duhem, a concept in physics is abstract when it is not faithful to the object it represents. But, as with any descriptions in physics, causal structures may be treated with more, or less, faithfulness. The real ruby laser that Maimon built had a particular concrete causal structure. But so too did the larger environment in which it ran that served as a reservoir to erase certain correlations in it. How realistically these structures can be copied in the model that the physicist reasons and writes equations about depends on the fit between reality and the mathematical needs of the theory; and even intricately detailed causal structures can be extremely abstract in Duhem's sense. Nevertheless, in a different sense, even a highly unrealistic object is more concrete with a causal structure than without one.

4. Causal Structures and Explanation

In his recent "Two Ideals of Explanation in Natural Science",[15] Ernan McMullin pursues a long-standing theme of his. He distinguishes between "retroductive", or "structural", explanations and "nomothetic" explanations. Nomothetic explanations are the empiricists' ideal: they explain an occurrence by showing it to be an instance of a general regularity. But, argues McMullin, in science the primary task is to explain not single happenings but recurring phenomena. We have found, from the nineteenth century onwards, that the most effective way to do so is via "a postulated structure of entities, relations, and processes". According to McMullin:

> Theory explains by suggesting what might bring about the explananda. It postulates entities, properties, processes, relations, themselves unobserved, which are held to be causally responsible for the empirical regularities to be observed.[16]

McMullin calls this second kind of explanation "retroductive . . . because it leads backwards from observed effect to postulated causes".[17]

In an earlier paper on the same theme, McMullin called these theoretical accounts "structural" explanations.[18] In a footnote to the newer paper, McMullin comments on his change in terminology: "In an earlier paper, I used the term 'structural' . . . to designate one common feature in the retroductive explanations of the natural sciences."[19] Presumably the common feature in question is the postulation of unobservable inner constitutions. What has been added in switching from structural to retroductive explanations? What has been added are the causal processes that link the inner structures with the phenomena they are supposed to explain. This addition provides the answers to two questions left open in the earlier paper: (1) Exactly how does McMullin's account differ from the commonly accepted deductive-nomological account of theoretical explanation articulated by Hempel? and (2) What is the explanatory nexus between the *explanans*, the structure, and the *explanandum*, a regularly recurring phenomena?

The two questions are related. An obvious answer to the second is suggested by Hempel's D.N. model: structures are explanatory in that they are governed by fundamental theoretical laws from which the "lower level" phenomenological laws in question can be deduced. In this case the answer to question (1) is that the accounts differ very little after all.

But this answer to question (2) is not one that McMullin would want to give. One of the major points of the second paper is to show that deductive subsumption under more general laws is no real explanation: "The inclusion of an explanandum under a regularity is not ordinarily sufficient to count as explanation in science. Someone notes that a piece of iron he is heating is expanding slightly. It will not help to inform him that this is what iron does."[20]

The example here is of a "single historical occurrence".[21] But exactly the same problem arises over regular occurrences and the "low level" generalizations that describe them. Whether an occurrence happens once or is general and repeatable, we need an account of why subsuming it under a generalization counts as explaining it. Hempel said that the subsumption showed that the phenomenon *was to be expected,* given the generalization. That answer has been heavily criticized by Bromberger,[22] Scriven,[23] Salmon,[24] and others and, rightly or wrongly, is not widely believed today. But no more acceptable alternative has taken its place.

McMullin's account differs importantly from Hempel's in that he downplays the role of laws, and the deductions one can make using them. For Mc-

Mullin, the laws are secondary to the structures; and he can maintain this view because for him the "explanatory nexus" is not subsumption but causation. The laws may play a role in helping us to trace the causal processes, but what explains phenomena in typical theoretical explanations are the structures and processes that bring them about.

A laser, we have argued, is characterized by its structure and its causal processes, unlike a reservoir, which is characterized only by its output. Following McMullin, we see that this is a critical difference. To add an account of an object's structure and causal mechanisms is not just to add some information or other, but is to add crucial explanatory information. In fact, it is to add just the information that constitutes explanation in modern theoretical sciences.

5. Why Causal Structure Gives Concreteness

There are two different intuitive ways that one object may be more concrete than another, and that we will combine in our account. The first is the simple idea that an object is more concrete if it has more properties. We don't know how to count properties. But we can still exploit this basic intuition by using the Aristotelian idea of nesting. If the properties used to describe one object are nested in those of another, the first description is less complete, and in that sense, more abstract than the other.

The second idea is that an object is more concrete when we know not just more about it, but more about what it really is. Again, we must face the problem of what counts as knowing more about what an object really is, and again we can take a solution from Aristotle. For Aristotle, the first and central role of a science is to provide explanatory accounts of the objects it studies and the facts about them. In each science the nature of the explanations may be different. In physics in particular, the basic objects, material substances are to be explained in terms of the four causes. Julius Moravcsik[25] suggests that we regard the four causes as different factors in an account explaining what a substance or a system really is: source (efficient cause), constituent (matter), structure (form), and perfection or function (final cause).

The fullest explanation will give as many factors as may be appropriate. An account which specifies all four factors will give all the relevant information for identifying the material substance and for understanding what it truly is. For example, if the account is to be of a particular person, the fullest account will mention all four factors, parentage, flesh and bones, human soul, and what an active human soul does, both in terms of the growth of person into a complete human, and the performance of a human soul. Other information is not relevant to understanding what this substance is as a person. Less information will be defective. Not all sciences, however, will use all explanatory factors: If we are to explain an Aristotelian unmoved mover, the constituent will be left out, because the unmoved mover is not encumbered with material constituents. Mathematics, on the other hand, is concerned with neither source nor function.

The four causes of Aristotle constitute a natural division of the sorts of properties that figure in explanation, and they thus provide a way of measuring the amount of explanatory information given. We can combine this with the idea of nesting to formulate a general criterion of abstractness: a) an object with explanatory factors specified is more concrete than one without explanatory features; and b) if the kinds of explanatory factors specified for one object are a subset of the kinds specified for a second, the second is more concrete than the first. Notice that, we are here nesting *kinds* of factors and not the individual factors themselves, and that, as with Aristotle's own hierarchy, the nesting provides only a partial ordering. For some pairs of objects, this criterion yields no decision as to which is more abstract. If the description of one object employs both form and function and the second only function or only causal structure, the first is more concrete than the second. But if one object is given just a causal structure and the second just a function, the criterion gives no verdict.

Consider a simple example. We might describe a paper cutter as an implement which rends paper. The definition refers just to the function of paper cutters. How a paper cutter rends paper is not a factor in the explanation. But the definition of scissors must give an account of the parts of scissors and how they cut, e.g., that they have at least two blades which operate by sliding past each other. We do not need to specify the material or constituent of scissors. Any material from which blades can be made will do. So too with the materials of lasers. Yet scissors are more concrete than paper cutters because more is specified about how the object works and performs its function.

Note that we could specify functions more finely. A smooth cutter will create smooth edges in rending paper. Smooth cutters are more abstract than scissors. For the variety of objects which can be scissors is restricted in a way in which smooth cutters are not. Their structural and material possibilities are still more diverse.

It might be objected that this notion of the abstract/concrete distinction is merely a variety of the general/specific distinction. But this is only partly the case. It is true that scissors are kinds of cutters, but it is not true that scissors are kinds of paper cutters, and much less that they are kinds of smooth paper cutters — some scissors cut meat. Yet we claim that paper cutters are more abstract than scissors.

A second objection is that one could very vaguely specify a structure and function for one item and specify a very specific function for another. For example, consider a mover where right and left are distinguished and an implement which cuts paper into frilled, square mats. Surely the cutter of frilled, square, paper mats is more concrete than the right-left mover, and surely this shows that we really ought to have given a general/specific distinction. What the example shows, rather, is that the general/specific distinction is relevant, and that our notion presupposes that the levels of description be not too divergent.

It is easy to see why this should be from the viewpoint of our distinction. As Aristotle observes about structure in the *Physics,* one cannot have just any matter with a given structure, but structure presupposes certain sorts of matter (one cannot make a laser of any material). So too a function presupposes certain structures (reservoirs have to be big). If the specification of structure is too general, as in the case of right-left movers, and we compare it with a function so specific, as that of cutters of frilled, square, paper mats, we may get the impression that the very general structure and function is less specific than the highly specific function and structures presupposed by it. In other words, the class of structures presupposed by the functional description of cutters of frilled, square, paper mats may be more specific than the structure mentioned in the description of right-left movers.

Let us now apply this criterion to the example of the laser and the reservoir. In the previous section we argued that in characterizing a laser we specify a complex causal structure and we specify what it does; in the case of a reservoir we specify only what it does. Although Aristotle's notion of form and end does not correspond perfectly with contemporary notions of causal structure and effect, we think the fit is good enough to adopt Aristotle's basic distinction to our examples in modern science: *form* as how the system operates and *function* as what the system does. Lasers then are more concrete than reservoirs because they are specified in terms of both form and function whereas reservoirs are specified in terms of function alone.

It is easiest to illustrate our concept of abstraction in the context of Aristotle's division of explanatory factors into four kinds; but this particular division is not necessary. Our basic view is that one system is more concrete than a second, then, when the kinds of explanatory features used to identify the second are nested in those that identify the first. So the general strategy requires only that we have a way of sorting explanatory features from non-explanatory features and of determining what kinds of explanatory features there are. It is a common assumption about modern physics, for example, that function is not an explanatory feature at all. If so, then structure and function is no more concrete than structure alone.

Even when a number of different kinds of explanatory features are admitted by a science, as with Aristotle's four causes, not all need be seen as equally central. McMullin, we have seen, argues that the causal structure is the primary explanatory feature in modern physics.

For Aristotle, too, structure plays a privileged role. It is the most important of the four explanatory factors in physics. The constituent is what a substance is made of; the function what it does or becomes; the source where it comes from. All these are a part of physical explanation. But structure is what the substance is. It is the essence of the substance, or what it is for this material body to be what it really is.

Structure, therefore, among all four of Aristotle's causes gives most fundamentally what a thing is, and hence by our criterion contributes most to making it concrete. The one factor that may seem equally important is matter.[26] For there is a sense in which matter—as that which individuates structure—is more concrete: it serves more to isolate the individual. But this is a different sense from the one we have been considering. For designating, or picking out, the individual does not tell what it is. This is primarily the job of the structure.

In the case of lasers and reservoirs the two relevant explanatory factors that we have isolated are function and structure. Between these two factors there is yet another reason for giving priority to causal structure in determining concreteness. For, given the causal structure, the function often comes along as a by-product; but not vice versa.

One cannot usually infer from effects to a causal history. At best one can describe classes of structures which would have such effects, only one of which could be the structure of the particular system at hand. But one can, at least in principle, infer from a description of the causal structure to what that causal structure does. This is not to say that one may not also infer all sorts of side effects. But all these effects are effects of that system as described. If a system is described in terms of causal structure, the addition of a description of certain effects of the system as described serves merely to isolate those effects we are interested in from among all the actual effects of the system as described. For this reason, descriptions of effects seem to add less to the descriptions of well-specified causal structures than do descriptions of causal structure to descriptions of well-specified causal effects. Hence, descriptions of specific causal structure are, in general, more concrete than descriptions of specific causal effects.

We see therefore that causal structure is privileged for a variety of reasons. From either McMullin's point of view, or the more traditional Aristotelian one, the causal structure plays the central explanatory role. It thus contributes most significantly in removing an object from the realm of the abstract, and making it more concrete. Consider the example of the crystal diode rectifier mentioned in the introduction. We can isolate three levels of concreteness here. The rectifier is identified by function alone; the diode rectifier by function and structure; and the crystal diode rectifier by function, structure, and matter.[27] Each is more concrete than the one before; but, we find intuitively, the first step, where causal structure is added to function, is far greater than the second, where the material is filled in.

This parallels the case of the laser and the reservoir. Adding causal structure to functional description, as in a laser, produces a critical change in the degree of concreteness of the object. We have seen this critical difference in the contrast between the laser and the reservoir.

Aristotle did not treat his four kinds of explanatory factors on a par. Structure or form supplies the essence, what the substance most truly is. If McMullin is right, modern science also gives central place to structural factors. Given the Aristotelian ideal that the explanatory factors are what tell us what an object really is, an object without structure — like a reservoir — will be a shadow object indeed.

Notes

1. Pierre Duhem, *The Aim and Structure of Physical Theory,* trans. Philip. P. Wiener (New York: Atheneum, 1962), pp. 155–156.

2. *Ibid.,* p. 151.

3. *Ibid.,* p. 152.

4. Ernan McMullin, "Two Ideals of Explanation in Natural Science", in H. Wettstein, ed., *Midwest Studies in Philosophy* (Minneapolis: University of Minnesota Press, forthcoming), p. 17 of manuscript.

5. Duhem, p. 166.

6. *Ibid.,* p. 169.

7. *Ibid.*

8. We learned to set Duhem in this historical context from Norton Wise. To learn

more about it, see his "The Flow Analogy to Electricity and Magnetism, Part I", *Archives for History of the Exact Sciences* 25 (1981): 19–70.

9. Nancy Cartwright, *How the Laws of Physics Lie* (Oxford: Oxford University Press, 1983), pp. 147–151.

The discussion in *How the Laws of Physics Lie,* as our discussion here, attempts to distinguish lasers from reservoirs in a way that is independent of the realism-anti-realism issue. But Cartwright did not there recognize the difference to be that between the abstract and the concrete. This only became clear in working with Mendell after the book was published. The emphasis on causal structure as the key to the distinction is missing in the book as well. There Cartwright stressed instead that the mathematical treatment of the reservoir was not motivated by any specific characteristics attributed to the reservoir in the model. This raises a further question: for a reasonably complete ["realistic" in the second sense discussed in *How the Laws of Physics Lie*] theoretical treatment, must the mathematics always be motivated by an understanding of the causal interaction or will other kinds of theoretical justification do?

10. A subclass of separate abstract entities would be those we know or conceive through extracting the property we are interested in away from the concrete body.

11. See M.-D. Phillippe. "*Abstraction, addition, séparation chez Aristote*", *Revue Thomiste* 48 (1948): 461–479.

12. Aristotle introduces his notion of the abstract as an *ancilla* to his theory of substance. It is not needed for the science of substance, which studies substances as they really are, nor for physics, which studies substance plus matter (in addition). Rather it is introduced solely to explain how we can have sciences and knowledge of non-substantial entities. But these sciences are the principal sciences of his time: arithmetic, harmonics, geometry, mathematical astronomy, mechanics, and optics. We do not need, however, to conceive of Aristotelian abstraction within the context of Aristotle's theory of substance. The notion is useful when we want to be able to distinguish between an object referred to in one way and the same object referred to in another way. We might say that such *de dicto* descriptions are abstract; *de re* descriptions, which do not focus on the attribute applied, are concrete.

13. For an example of this originally provided by Mendell, see Cartwright, p. 110.

14. L. Allen and D. G. C. Jones, *Principles of Gas Lasers* (New York: Plenum Press, 1967), p. 2.

15. McMullin, "Two Ideals . . .", p. 17.

16. *Ibid.,* p. 8.

17. *Ibid.*

18. Ernan McMullin, "Structural Explanation", *American Philosophical Quarterly* 15 (1978): 139–147.

19. McMullin, "Two Ideals . . .", p. 24.

20. *Ibid.,* p. 14.

21. *Ibid.*

22. S. Bromberger, "Why Questions" in R. Colodny, ed., *Mind and Cosmos* (Pittsburgh: University of Pittsburgh Press, 1966), pp. 86–111.

23. M. Scriven, "Causes, Connections, and Conditions in History" in William H. Dray, ed., *Philosophical Analysis and History* (New York: Harper & Row, 1966).

24. W. Salmon, *Statistical Explanation and Statistical Relevance* (Pittsburgh: University of Pittsburgh Press, 1971).

25. J. Moravcsik, "Aitia as Generative Factor in Aristotle's Philosophy", *Dialogue* XIV (1975): 622–638.

26. Here a more sophisticated view of matter may be appropriate. In his last thoughts on matter (Met. Z10–11), Aristotle wondered if the parts of the substance which occur in the definition which gives the essence could be matter. Matter as the locus of possibility and accident cannot occur in the essence or its definition. Hence, anything which appears in the definition which gives the essence, even a constituent of the substance, cannot be matter. We have here a way of distinguishing the Duhemian abstract and concrete. Matter will appear in the account of the concrete objects, but it cannot appear in the account of the abstract object. For that will only include essential features for the particular theory. On the other hand, we should not think that via Aristotle we can capture the notion of the Duhemian abstract object. For Aristotelian form or structure will be the structure of the concrete object, while the Duhemian abstract structure will not. In this respect Duhem is a Platonist.

27. Van Nostrand's *Scientific Encyclopedia* characterizes a crystal diode as "a diode consisting of a semi-conducting material such as germanium or silicon, as one electrode, and a fine wire 'whisker' resting on the semiconductor as the other electrode. Because of its low capacitance, the device finds considerable application as a rectifier or detector of microwave frequencies" (Princeton: D. van Nostrand Co., 1958), p. 519.

The Problem of
Indistinguishable Particles[*]

BAS C. VAN FRAASSEN

In the quantum-mechanical description of nature, an elementary particle is first of all characterized by some constant features, such as mass and charge. These features serve to classify them into basic kinds or types; physicists sometimes refer to particles characterized by the same constants as "identical particles". (In deference to philosophical usage I shall use 'identical' only in the strict sense in which no two distinct entities are identical.) In addition to these constant features, each particle is capable of various states of motion (represented by a Hilbert space). And that is all.

So if two particles are of the same kind, and have the same state of motion, nothing in the quantum-mechanical description distinguishes them. Yet this is possible. We have a dilemma: either this possibility violates the principle of identity of indiscernibles, or the quantum-mechanical description of nature is not complete. The dilemma could also be undercut: perhaps to conceive of such a particle as an individual, to which such a principle even could apply, is one of those many conceptual mistakes fostered by an upbringing in classical physics. A closer look very quickly reveals a whole cluster of problems, of which this dilemma is the center. Some sorts of particles obey the exclusion principle, and cannot have two in the same state of motion —but they too have been cited as a violation of identity of indiscernibles. Both sorts of particles exhibit (for reasons that appear to be related) statistical correlations in their behaviour which seem to defy causal explanation. And so forth.

In this paper I shall only try to identify, relate, and clarify the problems in this problem cluster. Though I will describe attempts at solution, including my own, I advocate none at this point. The very attempt to describe the problems systematically may reveal presuppositions whose denial could open up the way to a more satisfactory overall view.

153

1. Some Notes on the Literature

Elementary particles described by quantum mechanics fall into two classes distinguished by whether or not the exclusion principle applies. This is a principle concerning their aggregate rather than their individual behaviour; the principle is roughly formulated as ruling out occupancy of the same state by more than one particle of the given sort. There is a second division by aggregate behaviour: *fermions* are particles whose assemblies obey Fermi-Dirac statistics and *bosons,* those which obey Bose-Einstein statistics. The two divisions coincide: fermions are the particles to which the exclusion principle applies. The two dividing principles are not logically independent, but neither do they logically coincide. The first division is logically exhaustive, but other types of statistics exist (including the classical Maxwell-Boltzmann statistics, and various non-classical "parastatistics"). There is a third division, again factually, but not logically coincident with the first: fermions are the particles with half-integral spin, bosons the ones with integral spin.

Bose statistics and Pauli's exclusion principle were both introduced before the definitive formulation of either matrix or wave mechanics (Bose, 1924; Einstein, 1924 and 1925; Pauli, 1925). The quantum-mechanical treatment, properly speaking, of assemblies of identical particles was developed by Dirac (1926) and Heisenberg (1926); the general case of N particles and its relation to group theory was first presented by Wigner (1927). Central to this progress was the recognition of the permutation-invariance requirement for aggregate states of identical particles (see section 2 below).

The conceptual situation has continued to lead to debate and research in many areas; I list recent examples: in the foundations of physics (Aerts and Piron; Aerts), foundations of statistics (Sudarshan and Mehra; Costatini, Galavotti, and Rosa), philosophy of physics (Margenau; Reichenbach), quantum logic (Mittelstaedt), and general epistemology and metaphysics as related to science (e.g., the connected series by Cortes, Barnette, Ginsberg, and Teller). In a paper presented and commented on by Wesley Salmon in 1969 (but published in 1972), I addressed some of these issues, each of which has seen new contributions since then. There was a crucial aspect of Margenau's argument which I did not appreciate until after I had studied his and his students' writings on measurement, the Einstein-Podolski-Rosen paradox, and mixed states more closely (see section 3 below). We must be careful, I think, not to become overly pre-occupied with the principle of identity of indiscernibles in the discussions; yet Margenau was right to give it a central place, for it appears to relate to some aspect of each problem we shall encounter.

2. The Basic Invariance Requirement

Pauli's exclusion principle is generally presented in popular writings and textbooks as: no two electrons (more generally, fermions of the same sort) can be in the same state at once. The more basic invariance requirement, to which it was related by Fermi, Heisenberg, and Wigner, also entails that this is an inaccurate and sometimes misleading formulation.

Given two systems X and Y with states in Hilbert spaces H_1 and H_2, we represent the states of the composite system $(X+Y)$ in the tensor product space $H_{12} = H_1 \otimes H_2$. When X and Y are "identical" (of the same sort), and $H_1 = H_2 = H$ then $H_{12} = H \otimes H$. If X, Y are mutually isolated and respectively in the pure states φ, ψ, we assign $(X + Y)$ the product state $\varphi \otimes \psi$. In the identical case, $\psi \otimes \varphi$ would be another possible state, a permutation of the first one. More generally, the composite states take the form

$$\Phi = \Sigma c_{ij}\varphi_i \otimes \psi_j,$$

a permutation of which would be

$$\Phi' = \Sigma c_{ij}\psi_i \otimes \varphi_i.$$

States are in general not identical with, nor multiples of, their permutations. The question is whether the mathematical distinction reflects a real physical difference. That question appears to be answered in the negative by the basic postulate of

PERMUTATION INVARIANCE. *If Φ is the state of a composite system of identical particles, the expectation value of any observable A is the same for all permutations (i.e., $(\Phi, A\Phi) = (\Phi', A\Phi')$ for each observable A).*

Which vectors in $H \otimes H$ have this property? It can be deduced that they form two classes (henceforth omitting the symbol $\otimes$ when convenient):

Symmetric states: $\Phi = \Sigma c_{ij}(\varphi_i\psi_j + \psi_j\varphi_i)$
anti-symmetric states: $\Psi = \Sigma c_{ij}(\varphi_i\psi_j - \psi_j\varphi_i).$

If you substitute the symbols 'φ' and 'ψ' for each other, the symmetric state is left the same, while the antisymmetric state is multiplied by -1 (which also does not affect expectation values). It is easy to see that in the antisymmetric case, for each pair of indices i, j for which $\varphi_i = \psi_j$ we have a factor of *zero* (c_{ij} ($\varphi_i\varphi_i - \varphi_i\varphi_i) = 0$). This gives us the correct QM formulation of the exclusion principle. Call Φ in $H \otimes H$ an *exclusive product state* exactly if $\Phi = \varphi \otimes \psi$ with $\varphi \neq \psi$. Then we deduce:

(QM EXCLUSION PRINCIPLE) *Any anti-symmetric state is a superposition of exclusive product states.*

Returning now to the division of elementary particles, we note that they are classified by constants of motion (mass, charge, . . .). The quantum mechanical evolution operators described by Schroedinger's equation also do not turn symmetric states anti-symmetric, or conversely (the *symmetry-type* is also a constant of motion). So we have here in principle a further classificatory distinction, but in fact all known particles already fell into one class (all states symmetric, e.g., photons) or the other (all states anti-symmetric, e.g., electrons).

The permutations in question are permutations of the particles: intuitively the attribution of $\varphi \otimes \psi$ to $(X + Y)$ is that of $\psi \otimes \varphi$ to $(Y + X)$. Permutation invariance is therefore limited to empirical indistinguishability of the individuals. But this raises the *completeness question*: is there no real difference, or a real difference which quantum mechanics does not represent? This is just the sort of question raised by Einstein, Podolski, and Rosen, and (see below) is similarly related to non-classical correlations.

3. The Exclusion Principle and the Completeness Question

The idea of a connection with Leibniz's principle of the identity of indiscernibles (henceforth, PII) was perhaps first suggested by Herman Weyl, who referred to the exclusion principle as "the Pauli-Leibniz principle of exclusion". The nomenclature suggests the connection (which I proposed in my [1969], unaware of Weyl's earlier suggestion). Consider two distinct orbital electrons in an atom. Each is characterized by certain constants, definitive of electrons, plus a state of motion. The quantum-mechanical description admits nothing further, so if we assume that description to be complete, then the identity of indiscernibles requires their states to be different.

The assumption of completeness here may have a "metaphysical" air, but it is involved in the very application to atomic structure for which the exclusion principle was introduced. To show this, let us look at the application of this principle in the theory of atomic structure, and its reconstruction of the periodic table of chemical elements. For the structure of the hydrogen atom we introduce already three quantum numbers n, l, m, which together determine the hydrogen-atom wave functions. The *principal* number n determines the total energy E_n, and the number of nodes (radial and angular) which is $n - 1$. This number n can take any positive integral value. The *azimuthal* number l is the number of angular nodes; it is thus less than or equal to $n - 1$, and can otherwise take any non-negative integral value. It determines the square of the angular momentum; the *magnetic* quantum number m determines one component of angular momentum, which equals $m\hbar$. This quantum number takes the values 0, ± 1, ± 2, $\ldots$, $\pm l$. Considering only one electron, when $n = 1$ (electron in the lowest orbit), l and m are obviously constrained to be zero. Hence if these three numbers told us all, and the exclusion principle applied, there could only be one electron in the lowest orbit. But this is not so. In 1925 Goudsmit and Uhlenbeck introduced a fourth property to characterize the atomic electron, an intrinsic magnetic moment independent of its orbital motion, the *spin,* quantum number s being associated with total spin ($\frac{1}{2}$ for all electrons) and the quantum number m_s associated with one component thereof ($m_s = \pm \frac{1}{2}$). Having thus a new parameter with two possible values, we have at least two possible states for the case $n = 1$. If we now assume the description to be complete, there are exactly two possible states available, and if the exclusion principle is then applied, a maximum number of two electrons in the first orbit. This gets us as far as the model of the helium atom, with a nucleus of charge $+ 2e$ and two orbital electrons. Application to the three-atom lithium atom entails that the third electron cannot also have the lowest orbital state. For the second energy level ($n = 2$), there are four orbital states: as we have already seen l can then have value 0 and m value 0, or l have 1 and $m = 0$, ± 1. Adding that each of these states can be further distinguished by m_s, it follows that there are at least eight possible states available. Assuming completeness of this description, there are exactly eight states, and applying the exclusion principle, we conclude that there can be at most eight electrons in the second orbit. And so forth. So the completeness assumption, used above in the suggested deduction of the exclusion principle, is also essential to its primary application in atomic theory, and not extraneous to the scientific context.

In the light of the above, it is with some surprise that we find Margenau (in 1944 and again in 1950) citing the electron as violating Leibniz's principle, in writings specifically devoted to the exclusion principle. But looking more closely at the permutation invariance principle, I believe that we can reconstruct an argument that leads to Margenau's conclusion (which I did not appreciate when I wrote my [1969]). Consider a two-particle system in the simple anti-symmetric state $\varphi_{12} = \varphi_1 \otimes \varphi_2 - \varphi_2 \otimes \varphi_1$. What states, if any, can we attribute to the individual components?

Well, suppose I make a measurement of observable A on one component. This is equivalent to measuring $A \otimes I$ (with I the identity operator) on the composite system. Calculating the expectation value, we find exactly the same answer as for the case of measuring A on a single system in a mixed state, which is a half-and-half mixture of φ_1 and φ_2. From other discussions by Margenau and his students (of the measurement problem and the quantum mechanical paradoxes) we know that in these circumstances they conclude that the attribution of a pure state to the component system is not only unwarranted, but *false*. Instead they assign each particle that same mixed state. There are strong, though not entirely uncontroversial, consistency arguments for their conclusion.

The view at issue here is the so-called "ignorance interpretation" of mixed states. It is true of course that if I am sure that a given particle was prepared either in state φ or in state ψ, but I have no idea which, I can adequately represent the situation by a mixed state, a half-and-half mixture of these two pure states. In that case my ascription of a mixed state reflects my ignorance. But mixed states are encountered also in a different context. Sometimes a complex system has a pure state, and it is different from any pure state we would ascribe to it, in view say of the spatial separation of its components, on the basis of any supposition of pure states for those components. This happens typically after past interactions; Schroedinger called it *the* distinguishing feature of quantum mechanics (in a paper related to the Einstein-Podolski-Rosen paradox). In that case however, predictions about observables relating to *one* component can be based on a "reduction of the density matrix", which ascribes a mixed state to the component part. This ascription is *ex hypothesi* incompatible with the idea that the mixed state is a mixture of pure states that the component may have for all we know. One view is that the component has no state at all (of its own). Another view, which I attribute here to Margenau, is that the mixture is the state of the component; and correlatively, that mixtures *are* the possible states of motion, with pure states representing only an unprivileged special case.

But on this view, the two particles discussed in the second last paragraph (which are as an aggregate in a superposition of exclusive product states) are not themselves in different states at all. Hence they are literally indiscernible —though not identical.

It is clear that the crucial step in Margenau's argument about the exclusion principle must consist in a stronger completeness claim for quantum-mechanical description—one that rules out the universal applicability of an "ignorance" interpretation of mixed states. Could the stronger claim be independently disputed? A careful inquiry into the nuances of the completeness question is needed. Or could the very individuality of the component systems be denied in such a fashion that PII becomes inapplicable?

One possible reaction (I won't put it more strongly) is to consider the weakened completeness claim to be found in modal interpretations of quantum mechanics. These interpretations distinguish between *states* and *events*—

an event happening exactly when a (non-constant) observable has some definite value. The state is incomplete in that it gives only probabilistic information about events, but complete in that no other information has any predictive value for future events. In the most restrictive version (Copenhagen version) of the modal representation, a pure state is actually complete in both respects, but mixed states are not. (Thus at the end of the measurement, the apparatus is in a mixture of pointer-reading "states", but the pointer is *actually* at one specific number on the dial.) At least formally, we have here a resolution of the tensions between the identity of indiscernibles and the exclusion principle. If it is held that the fermions have no features left undescribed by the quantum mechanical formalism, this interpretation implies that a pair of fermions cannot be in a product of two identical pure states, but each could be in the same mixed state (while *actually* subject to different ones of the possible events allowed by that state).

4. Considerations of Genidentity for Bosons

Bosons are particles capable only of symmetric aggregate states; they too have been suggested as examples of numerically distinct entities that may be indiscernible. There has appeared a connected series of articles dealing with this: Reichenbach (1956, section 26), van Fraassen (1969), Cortes, Barnette, Ginsberg, and Teller; all but the first referring to the preceding one.

Bose's introduction of his statistics was the last step in a historical development directly concerned with electromagnetic radiation and statistical mechanical analogies. If a certain amount of light, say, is introduced into an evacuated enclosure with perfectly reflecting walls, we have a situation in some ways similar to an enclosed body of gas. Specifically, the "radiation gas" exerts a pressure on the walls and work must be expended to decrease the volume. If now a piece of matter is introduced, capable of emitting radiation in every frequency, then emission and absorption will happen until their two rates are equal, and remain so: an equilibrium is reached. Experiments suggested that in this equilibrium situation, the intensity of light of a given frequency in the enclosure is a function solely of that frequency and the temperature of that enclosure. The description of that function is exactly the subject of Stefan's, Wien's, Rayleigh's, and finally Planck's laws of radiation. While Stefan's law is based on experimental results, and was accepted as a partial constraint on the required function, Wien's and Rayleigh's were based respectively on a thermodynamical argument and a deduction from the classical laws of electromagnetism (both using additional assumptions). These latter two turned out to be erroneous on the whole though approximately correct in certain limits. It was exactly at this point that Planck introduced his "quantum theory", and was able to deduce his empirically satisfactory radiation law. But the deduction was based partly on classical assumptions and partly on assumptions incompatible with classical physics, not a theoretically satisfactory situation.

Einstein's treatment of the photo-electric effect, in which corpuscular properties were attributed to energy quanta (radiation of frequency v consisting of photons having energy hv and momentum hv/c) made the statistical mechanical view more than a mere analogy. The pressure which the radiation exerts on the walls can now be attributed to the impact of the photons, exactly the same mechanism as for an ordinary gas. If we now apply Boltzmann's classical statistical mechanics to the distribution of numbers of photons over the various energy levels (corresponding to intensities of radiation over various frequencies) for an equilibrium situation, we obtain Wien's law. But Planck's law should result. Hence Bose introduced a non-classical assumption of equiprobability. The classical assumption would be that each arrangement of individual particles, classed together when they have the same energy level, is equiprobable. Bose's assumption was that the identity of the particles is to be ignored, and each possible assignment of occupation numbers to the different energy levels is equiprobable. This was *ad hoc*: it led to Planck's law. But in retrospect, the quantum-mechanical treatment justified the relevant dismissal of individuality: the correct division into equiprobable cases is obtained if we assume all possible aggregate states of the "photon-gas" to be equiprobable, but suppose that these are symmetrical composite states. (The correct linkage between the Bose-Einstein treatment of radiation and symmetric wave-functions is indicated by Wigner [1926, page 495].) One main suggestion recurring in the literature is that the lack of individuality or identifiability of the bosons thus appealed to in the usual explanation of Bose's statistics has to do with identity across time.

The traditional questions concerning identity through time were reformulated precisely by Reichenbach, first in connection with relativity (1928, section 43) then for quantum mechanics (1956, section 26). He uses the classical particle/wave distinction as illustration. A floating cork bobs up and down when a wave reaches it; thus we see that no water moves laterally, although the wave moves across the surface. If the individual water droplets, or better its molecules are entities persisting in time, the wave is merely a changing configuration of these entities. In Reichenbach, a particle has *material identity* (its temporal stages, or the events involving it, are *genidentical* with each other) and the wave does not. Hence questions of individuation or identification of waves (which may form superpositions) are either misplaced, or settleable by convention.

Even in the context of the classical world picture, this distinction may be challenged. Are groupings of events into individual histories, by contiguity and succession, anything more than just that—more than a conventional if practically important classification? Reichenbach's answer was, in part, that one grouping may give us a world subject to, and another a world devoid of, causal anomalies. This distinction, he thought, could bestow objectivity on the genidentity relation. In the case of the quantum-mechanical world, it was not clear (and now seems entirely unlikely) that any grouping of events into individual histories will eliminate all causal anomalies in his sense. (His precise form of the problem was that neither a classical particle nor a classical wave picture will by itself fit all the phenomena.) But at least it was possible to conclude, according to him, that to regard bosons as entities persisting through time (having material identity) entails causal anomalies.

The causal anomaly to which he points is the statistical correlation in boson behaviour, even in the absence of perturbing forces. This correlation I shall discuss in the next section. Let us here just address the suggestion that bosons are "not genidentical". This means that where intuitively we have, say, an assembly of n photons each persisting in time, we really have only at each moment n photon-stages (temporal slices), and there is no objectivity of any sort to the classification of one of these photon-stages at time t belonging to the same photon as one or other of the stages at time $t + d$. A photon-stage at a certain time is really no more than an event—the *being-occupied* of a certain photon-state. But now we recall that the boson aggregate states are symmetric, which entails that several or even all of them may be in the same pure state at once. ($\varphi \otimes \varphi$ is a symmetric two-particle state.) Hence all n events may have exactly the same character—there are n being-occupieds of the same photon state. Not being individuated by historical connections to prior stages, and the quantum mechanical state being assumed to give a complete description, we conclude that we have here to do with true numerically distinct indiscernibles. Hence PII is violated (a conclusion which Reichenbach does not draw). Were PII to operate, non-genidentity would entail that there is no multiple occupancy of boson states.

In the next two sections I shall discuss the statistical correlations and take issue with Reichenbach's views on causality. For now let us note that if we accept PII, then for the case even of instantaneous particle-stages we have a dilemma: either there cannot be more than one of exactly the same sort (as characterized by the quantum-mechanical description) or else they are really distinguished by some (non-quantum-mechanical) "hidden factor" (which may be genidentity, or something else—e.g., perhaps the particle-stage is a configuration of some underlying medium whose parts are separately individuated, as for classical waves). If we ignore other possible hidden variables, then it is exactly the non-genidentical entities which must obey the exclusion principle, according to PII, and the genidentical ones which need not. Presence and absence of material identity through time could therefore "explain" what makes a particle a fermion or a boson.

This inversion of Reichenbach's classification, which I described in my (1969), was challenged by Cortes, who argued that it is better to reject PII than to accept as real an empirically vacuous hidden factor. Barnette accused Cortes of confusing metaphysics and epistemology; Ginsberg showed that in quantum field theory Barnette's reasoning looked much less plausible; and Teller argued cogently that these discussions left a number of unsolved problems. (Aerts and Piron, I should add, take exactly the view that bosons are distinguished by some feature ignored in the physical description—see Aerts, page 402.) I note here that, at a crucial point in the preceding paragraph, a completeness claim occurred again; resolution here probably depends on the issues of the preceding section. Again we must also ask: could we perhaps deny the individuality of component systems in a much more radical way (beyond denial of genidentity) so as to dissolve the problem? (But for bosons, this denial of individuality would have to apply even to an assembly each of whose members is in a *pure* state.)

5. Theoretical Unification of the Different Statistics

The classical Maxwell-Boltzmann and the non-classical Bose-Einstein and Fermi-Dirac statistics can (and are) also studied abstractly, independent of their physical basis. These studies look disconcertingly classical, but Sudarshan and Mehra show that Bose and Fermi statistics, with their major physical consequences, can be formulated for a classical phase-space with a preferred cell-size (e.g., h^3) as the only non-classical feature. To give a feeling for what they are like, consider the case of the two individuals distributed over two cells (which I shall call H and T, for "heads" and "tails", but these names imply nothing about them).

	H	T	p^{be}	p^{fd}	p^{mb}
Case 1	a_1, a_2		$1/3$	0	$1/4$
Case 2	a_1	a_2	$1/6$	$1/2$	$1/4$
Case 3	a_2	a_1	$1/6$	$1/2$	$1/4$
Case 4		a_1, a_2	$1/3$	0	$1/4$

Here p^{mb} is the familiar equiprobability assignment to all *logically* possible cases. The others may be understood thus: p^{fd} obeys the exclusion principle (zero probability for multiple occupancy) and p^{be} treats as equiprobable all numerical distributions (two in the first cell; one in each cell; two in the second cell). The symbols a_1, a_2 may of course be no more than indices of the sort we used in description of quantum-mechanical states.

All three obey the condition that permutation of these indices does not change probabilities: isomorphic cases 2 and 3 are always assigned the same probability. (Carnap called this "*symmetry*".) We also see that on the case of *one* individual all three agree: the proposition that a_1 is in H is the disjunction of cases 1 and 2, to which each of them gives the sum $1/2$. Thus the differences concern aggregate behaviour. Can we give a unified account of them — an account that places each in a systematic classification, and may throw some light on the basic physical differences they model?

The most far-reaching recent studies in this abstract vein are undoubtedly those by Costatini and his colleagues (1979, 1982, 1983). For the case k cells and n individuals, they define the characteristic *relevance quotient*

$$e\ (p) = \frac{p\ (a \text{ is in cell } Vi \mid b \text{ is in cell } Vj)}{p(a \text{ is in cell } Vi)} \quad \begin{array}{l} \text{for } i \neq j; \\ a \neq b \end{array}$$

which they prove to be well-defined (i.e., the same for all a, b, i, j) on the basis of certain general conditions satisfied by the three statistics. They then show that the three statistics are uniquely differentiated by the value of e:

$$e(p^{mb}) = 1$$
$$e(p^{be}) = k/(k + 1)$$
$$e(p^{fd}) = k/(k - 1).$$

This is a precise measure of the correlation of individual behaviour modelled by the three statistics. In our table we can quickly verify: $p^{mb}(Ha_1 \mid Ta_2) = p^{mb}(Ha_1)$ so $e(p^{mb}) = 1$; $p^{be}(Ha_1 \mid Ta_2) = (1/3)$ so $e(p^{be}) = (1/3) \div (1/2) = (2/3)$; $p^{fd}(Ha_1 \mid Ta_2) = 1$ so $e(p^{fd}) = 1 \div (1/2) = (2/1)$. The case $e = 1$ is that of total statistical independence between the individuals. We shall go further into this topic of correlation in the next section. For now, note however that their systematic classification has room for other statistics ("parastatistics") characterized by other relevance quotients. Hence it does not illuminate why only those three cases should appear in physical situations.

The above results were presented in Carnap's framework for probability theory, in one of their publications (1982). In my (1969) I also characterized the three statistics in the terms of that framework, though proposing a unification of a rather different sort. To explain it I must say something about Carnap's program. He took it that just about any assignment of probabilities could be correct in some possible situation, but also thought that this assignment would be the result of conditionalizing a certain basic "ur-probability", a logically determined prior probability function, on a set of propositions characterizing that situation. (Since his interest was in confirmation theory, the situations were epistemic, and the characteristic set of propositions exactly the known data or information given.) The properties that single out the ur-function, such as *symmetry* (see above) and *regularity* (assignment of *zero* only to logically impossible propositions) are not preserved under conditionalization. (For example, the information that a coin has landed heads up destroys symmetry [since this coin is now distinguished from other coins] and regularity [because the contingent proposition that it landed tails up is now assigned *zero*].) Hence these properties do not generally characterize the correct probability assignment for a particular sort of situation.

Could we find an ur-probability such that p^{mb}, p^{be}, and p^{fd} are all conditionalizations of it? In fact, Carnap himself proposed p^{be} (his m^*) as *the* candidate (relative to his program) for the ur-probability (though later he revised this opinion). Let us try it as a candidate for the ur-probability for assemblies of physical particles.

For Carnap, the cells are characterized by families of predicates—thus in the study of an urn-problem, the predicates might be 'cubical' and 'red', the cells being cubical-red, cubical-nonred, noncubical-red, and noncubical-nonred. The four complex predicates representing the cells are called Q-*predicates* (logically strongest consistent predicates in the language). Let us say that a family of predicates *individuates* a set of individuals if no two of them can be alike with respect to all these predicates (i.e., no Q-predicate formed from this family can characterize more than one such individual). We now have three possible situations:

(1) The family does not individuate the individuals (every logically possible state-description can be true).

(2) The family as a whole individuates (only those state-descriptions in

which each individual satisfies a different Q-predicate can be true).

(3) A proper sub-family individuates (each individual satisfies a different Q-predicate *of that subfamily*, in each state-description that can be true).

Cases 2 and 3 can hold only relative to some postulates, on which the ur-probability is to be conditionalized, for *logically* speaking of course every state-description could be true. In fact, it is easy to formulate the relevant postulates. Let $a_1, \ldots, a_n$ be individual constants and let $\{F_1, \ldots, F_k, G_1, \ldots, G_m\}$ be the total family of predicates. Let $\{Q_1, \ldots, Q_q\}$ with $q = 2^{k+m}$ be the set of Q-predicates for the whole family and $\{Q'_1, \ldots, Q'_r\}$ with $r = 2^m$ the set of Q-predicates for $\{G_1, \ldots, G_m\}$ alone. Then those postulates are

Situation 2: $Q_h(a_i) \supset \sim Q_h(a_j)$ for $h = 1, \ldots, q$

Situation 3: $Q'_h(a_i) \supset \sim Q'_h(a_j)$ for $h = 1, \ldots, r$

in both cases for each $i \neq j$ from 1 to n.

In the table given at the beginning of this section it is already illustrated that p^{fd} is p^{be} conditionalized on the situation 2 postulates. That is, with H, T as the two Q-predicates, we have

$$p^{fd}(--) = p^{be}(-- \ Ha_1 \supset Ta_2 \ . \ \& \ . \ Ha_2 \supset Ta_1).$$

This is easily checked by noting that the odds between the *remaining* state-descriptions are the same in both cases ($\frac{1}{6} \div \frac{1}{6}$, $\frac{1}{2} \div \frac{1}{2}$). This is a trivial case but the argument is general: the situation 2 postulates rule out all *structure-descriptions* exhibiting multiple occupancy of cells. The remaining structure descriptions each contain the same number of state-descriptions, so all remaining state-descriptions (as well as, separately considered, all remaining structure-descriptions) are treated as equi-probable by p^{be} — just as p^{fd} does.

To illustrate the effect of the situation 3 postulates we need a bigger table. Let $k = m = 1$, so we have only four cells (and G_1, $\sim G_1$ are the Q-predicates of the relevant subfamily), and let $n = 2$.

	F_1	$\sim F_1$	G_1	$\sim G_1$	
Case 1	a_1, a_2	–	a_1	a_2	$\frac{1}{8}$
Case 2	a_1	a_2	a_1	a_2	$\frac{1}{8}$
Case 3	a_2	a_1	a_1	a_2	$\frac{1}{8}$
Case 4	–	a_1, a_2	a_1	a_2	$\frac{1}{8}$
Case 5	a_1, a_2	–	a_2	a_1	$\frac{1}{8}$
Case 6	a_1	a_2	a_2	a_1	$\frac{1}{8}$
Case 7	a_2	a_1	a_2	a_1	$\frac{1}{8}$
Case 8	–	a_1, a_2	a_2	a_1	$\frac{1}{8}$

Here cases 1–4 are mutually non-isomorphic; case i is isomorphic to case $4 + i$ (for $i = 1, 2, 3, 4$). All other cases are non-isomorphic to these and ruled out by the situation 3 postulates. So we have here four of the original structure descriptions, in fact. By p^{be} all structure-descriptions were equiprobable, and conditionalization leaves the "internal odds" the same, so the remaining four are now still equiprobable (¼ each). Now we notice that the *state-descriptions* in the *other* subfamily $\{F_1\}$ correspond to those remaining structure descriptions in the whole family $\{F_1, G_1\}$, and hence are equiprobable. But that means that p^{be}, so conditionalized, *coincides with p^{mb} on the other subfamily* (on the remainder of the overall family). An example is: we have two coins (a_1, a_2) one of which is scratched (G_1) and the other not; each can be heads (F_1) or tails ($\sim F_1$) independently of the other. If we now look at heads *vs.* tails *alone*, p^{be} conditionalized on the relevant postulate ($G_1 a_1 \supset \sim G_1 a_2$) gives us the effect of p^{mb}.

To sum this up then, we can see the three statistics as special cases of the same principle (prior equi-probability for structure descriptions) for situations of different extent of individuation by the predicates considered. Again we see here the relevance of completeness and the PII. A claim of individuation is a completeness claim for a family of predicates; the PII entails that there must always be some family of predicates which individuate. Thus in this perspective, situation 1, the boson case, provides the challenge to PII.

6. Causality and Correlation

Viewed from a classical perspective, the Bose and Fermi statistics entail correlations which cry out for causal explanation. This is made clear by Costatini's relevance quotient classification, but let us begin with the corresponding physical considerations. The non-classical correlations are easily illustrated by the quantum mechanical model (briefly considered by Margenau) for two identical particles moving in parallel (say, along the x-axis) with sharp velocities u and v.

In this simple illustration I will not normalize the state-vectors; hence the value $f(x,x')$ derived is *proportional* to the probability that the two particles are found at positions x and x'. We are given two particles in momentum eigenstates, with motion along the x-axis. Treated individually, they would be assigned states $\psi_1 = e^{iux}$ and $\psi_2 = e^{iux'}$; setting $p = (ux + vx')$ and $q = (ux' + vx)$

and choosing first the anti-symmetric case, we assign the composite system the state

$$\psi_{12} = e^{ip} - e^{iq}.$$

Applying the Born rule for probabilities of position, we must evaluate

$$
\begin{aligned}
\psi_{12}{}^* \psi_{12} &= \left(e^{ip} - e^{iq}\right)^* \left(e^{ip} - e^{iq}\right) \\
&= \left(e^{-ip} - e^{-iq}\right)\left(e^{ip} - e^{iq}\right) \\
&= 2 - e^{i(q-p)} - e^{-i(q-p)} \\
&= 2 - (\cos(q-p) + i\sin(q-p)) \\
&\quad - (\cos(q-p) - i\sin(q-p)) \\
&= 2[1 - \cos(q-p)] = 2\ [1 - \cos(ux' + vx - vx' - ux)] \\
&= 2[1 - \cos((v-u)(x-x'))].
\end{aligned}
$$

Hence the probability $f(x,x')$ is proportional to $[1 - \cos((v - u)(x - x'))]$.

Taking secondly the symmetric case, we assign state $\psi'_{12} = e^{ip} + e^{iq}$ and evaluate

$$
\begin{aligned}
\psi_{12}{}^* \psi'_{12} &= \left(e^{ip} + e^{iq}\right)^* \left(e^{ip} + e^{iq}\right) \\
&= \left(e^{-ip} + e^{-iq}\right)\left(e^{ip} + e^{iq}\right) \\
&= 2 + e^{i(q-p)} + e^{-i(q-p)} \\
&= 2\ [1 + \cos(q-p)] \\
&= 2\ [1 + \cos((u-v)(x-x'))]
\end{aligned}
$$

and hence the probability $f'(x,x')$ is in this case proportional to $[1 + \cos(u-v)(x-y)]$.

Obviously we get *zero* for $a_1 = a_2$ and positive probability values increasing with d (up to a point) for $a_1 = a_2 + d$, in the antisymmetric case. Thus it looks as if the two particles repel each other, and do so the more if their velocities are more nearly the same. In the symmetric case, the probability is always positive, but increases as we take a_1, a_2 closer together. Thus it looks as if they attract each other. The absence of forces of attraction and repulsion (though there were speculations about "Pauli forces", and the conscious metaphor of "exchange forces") makes one want to say that the particles seem to know each other's state, and either shy away from it (fermions) or try to follow suit (bosons).

The abstract formulation of the statistics show how they model these correlations in their basic equi-probability divisions. In the following table, each structure description is represented by a single comprised state-description, with the isomorphic ones indicated by names ('S_i') only. The numbers are assignments of probability to the *structure* descriptions.

p^{fd}	p^{be}	p^{mb}	F_1	F_2	F_3	F_4	isomorphic state-descriptions	
⅙	⅒	2/16	a	b			S_1	S_2
⅙	⅒	2/16	a		b		S_3	S_4
⅙	⅒	2/16	a			b	S_5	S_6
⅙	⅒	2/16		a	b		S_7	S_8
⅙	⅒	2/16		a		b	S_9	S_{10}
⅙	⅒	2/16			a	b	S_{11}	S_{12}
0	⅒	1/16	a,b				S_{13}	
0	⅒	1/16		a,b			S_{14}	
0	⅒	1/16			a,b		S_{15}	
0	⅒	1/16				a,b	S_{16}	

p^{fd}, p^{be}, p^{mb} each assign ¼ to the event F_1a (a in cell F_1), i.e., to the class of state-descriptions $\{S_1, S_3, S_5, S_{13}\}$. The conditional probabilities are not the same:

p^{fd} $(F_1a \mid F_3b) = \frac{1}{3}$ positive correlations ("repulsion")

p^{be} $(F_1a \mid F_3b) = \frac{1}{5}$ negative correlation ("attraction")

p^{be} $(F_1a \mid F_1b) = \frac{2}{5}$ positive correlation

p^{mb} $(F_1a \mid F_3b) = \frac{1}{4}$ no correlation

Reichenbach had proposed that every genuine (persistent, resilient) positive correlation must have a causal explanation. The quantum mechanical description does not bear this out. The challenge quantum mechanics presents here (correlations that fit no causal model) is therefore fundamentally the same as in the EPR and Bell Inequality cases. If the three statistics can be unified (with p^{be} as basic) in the way suggested above, we may have a way of reconciling our intuitions. The intuitive idea would be that correlations "built into" the basic statistics require no explanation, but only divergences therefrom. To make this precise we need some measure of such divergence (Jaynes-Kullback relative information is a *prima facie* measure, as are some functions described by I. J. Good), and I think also a more liberal criterion of explanation than Reichenbach's (something like redistribution of odds that "tells for" the feature to be explained). But these are at the moment only tentative suggestions.

7. Naming and Describing in Quantum Logic

The assertion that at least two photons were emitted during a certain interval by a certain atom is at first blush easily formulable. The formula will have the form '$(Ex)(Ey)(Fx \ \& \ Fy \ \& \ x \neq y)$'. Semantic analysis of the usual sort entails that this formula is true exactly if there are entities such that if one is (momentarily) the referent of 'x' and the other of 'y', then '$Fx \ \& \ Fy \ \& \ x \neq y$' is true. But what exactly is required for an entity to be the referent of 'x'? At one extreme we have what Putnam contemptuously calls the *magical theory of reference*. It says in effect that really nothing is required: if there are two individuals, there also exist functions mapping the set 'x', 'y' into this couple of individuals. The semantic analysis can be taken as saying simply that there exists some function f such that if we regard 'x' as denoting $f('x')$ and 'y' as denoting $f('y')$ then '$Fx \ \& \ Fy \ \& \ x \neq y$' is true.

This attitude may or may not suffice as long as we look only at such quantificationally closed sentences. It becomes vastly more precarious if we try to extend it from momentary, arbitrary naming by free variables to reference by fullfledged individual terms such as names or descriptions. At the other extreme we find the *causal theory of reference* which requires that some causal chain connects the (production or use of the) term and its referent. However obscure the notion of causal chain may be, this view of reference would seem to preclude differential naming of two photons in the same state — since entering into distinct causal chains would surely distinguish them in a way that quantum mechanics does not recognize. Unless of course the notion of causality be sufficiently metaphysical, or the quantum mechanical description of nature sufficiently incomplete!

It is not surprising therefore that the problem of indistinguishable particles has recently begun to fascinate writers in the area of quantum logic (see especially Mittelstaedt; Dalla Chiara and Toraldo di Francia). This is the youngest of our problem areas. I can do no better at this point than to refer to the fact that some recent approaches to quantification and singular terms are much less sensitive than others to application problems for standard semantic concepts. There is first the theory of quantification, names, and identity for complete lattices in terms of abstractors (van Fraassen 1982). Second, and perhaps less abstruse, there is the general notion of quantifiers suggested by Kit Fine and elaborated by Charles Daniels. On this second view, a quantifier is based on a function which takes propositions into numbers. Let 'A' be a wff, 'v' a variable, 'X' a proposition; then $V('vA')$ is a function f such that $f(X) = n$ just if there are exactly n valuations V' differing from V at most at v such that $V'('A') = X$. Then the formula 'NvA' is true at world w exactly when 'N' denotes the number $\Sigma\{V('vA')(X):w \in X\}$. I know this looks rather complicated at first; but less so if you assume the lattice of propositions to be finite and take most of the summed numbers to be zero. In any case, the manoeuvre is suggestive of the idea that although in a permutation-invariant state we do not treat the particles as inherently distinguishable, *number* remains a well-defined observable.

8. Instead of a Conclusion

Although I have not been entirely neutral in my exposition, and have suggested some approaches to specific problems in the cluster, I was quite sincere in my initial announcement that I have no general or overall solution to offer. Unfortunately, there are too many loose ends to pretend otherwise, even if we suppose that all my favourite approaches will turn out to work.

When it comes to the interpretation of quantum mechanics I do think, despite the many problems and disagreements, that there has been a great deal of progress in recent years (or at least, recent decades). Foundational studies of many sorts have provided us with a much deeper understanding of the structure of the theory. As a result alternatives have been sharpened and their consequences clarified. If we now disagree, for example, on the ignorance interpretation of mixtures, or the (relative) completeness of quantum mechanics, we can all quickly locate exact problem areas and consistency questions for our views. Most important perhaps has been the progressive shift of inquiry into facets of aggregate behaviour and the structure of composite systems. Revolutionary as the introduction of indeterminism and discreteness and absence of joint distributions were, the most radical features of the quantum-mechanical world are undoubtedly those pertaining to wholeness—the challenge to the ingrained idea that what is distinguishable is separable is separate. (This is also the theme of the conclusion of the recent paper by Dalla Chiara and Toraldo di Francia—a theme which goes back of course to Bohr's replies to Einstein, but which returns with devastating new impact at every new turn in our *problematique*.) Addressing issues in the foundations of quantum statistics, which by definition goes beyond the theory of the individual system, we take this shift one step further in the same direction.

Note

*It is a great honor and a pleasure to contribute to a volume dedicated to Ernan McMullin, who brings to our field an unrivalled breadth of learning in both philosophy and science, and great philosophical insight. The subject on which we have our farthest-reaching disagreement is undoubtedly scientific realism. Rather than write on that subject, I chose a topic in the philosophical foundations of quantum mechanics, exactly because I consider Professor McMullin to be perhaps, among scientific realists, the most sensitive to the almost inconceivable and strangely paradoxical differences between the micro-world described by that theory and the macro-world with which we are (or took ourselves to be) familiar. On the search for an answer to one question scientific realists and anti-realists collaborate rather than compete: how could the world *possibly* be the way physical theory says it is?

References

Aerts, D. 1981. "Description of Compound Physical Systems", pp. 381–404 in E. Beltrametti and B. van Fraassen, eds., *Current Issues in Quantum Logic*. New York: Plenum.

Aerts, D., and Piron, C. 1979. "Physical Justification for Using the Anti-symmetric Tensor Product", preprint TENA. Free University of Brussels.

Barnette, R. L. 1978. "Does Quantum Mechanics Disprove the Principle of the Identity of Indiscernibles?". *Philosophy of Science* 45: 466–470.

Bose, S. N. 1924. "Plancks Gesetz und Lichtquanten-hypothese". *Z. Physik* 26: 178–181.

Bub, J. 1973. "On the Completeness of Quantum Mechanics", pp. 1–65 in C. A. Hooker, ed., *Contemporary Research in the Foundations and Philosophy of Quantum Theory*. Dordrecht: Reidel.

Carnap, R. 1950. *Logical Foundations of Probability*. Chicago: University of Chicago Press.

Cartwright, N. 1983. *How the Laws of Physics Lie,* chapter 9. Oxford: Oxford University Press.

Cortes, A. 1983. "Leibniz' Principle of the Identity of Indiscernibles: A False Principle". *Philosophy of Science* 43: 491–505.

Costatini, D. 1979. "The Relevance Quotient". *Erkenntnis* 14: 149–157.

Costatini, D., Galavotti, M-C., and Rosa, R. 1982. "A Rational Reconstruction of Elementary Particle Statistics". *Scientia* 117: 151–159.

———. 1983. "A Set of 'Ground Hypotheses' for Elementary Particle Statistics". *Il Nuovo Cimento* 74B, N.2: 151–158.

Dalla Chiara, M-L., and Toraldo di Francia, G. 1983. "Individuals, Kinds and Names in Physics" forthcoming in *Logica e Filosofia della Scienza, Oggi.* Proceedings Soc. Italiana di Logica e Filos. della Scienza, San Gimignano, 1983.

Daniels, Ch. 1980. "Towards an Ontology of Numbers". MS.

Dirac, P. 1926. "On the Theory of Quantum Mechanics". *Proc. Royal Soc. London,* Ser. A 112: 661–677.

Einstein, A. 1924/1925. "Quantentheorie des einatomigen idealen Gases". *Preussische Akad. der Wissenschaften (Phys.-math. Klasse) Sitzungsberichte* (= Berliner Berichte), 1924: 261–267; and 1925: 3–14.

Fermi, E. 1926. "Zur Quantelung des idealen einatomigen Gases". *Z. Physik* 36: 902–912.

Fine, A. 1982. "Antinomies of Entanglement: The Puzzling Case of the Tangled Statistics". *Journal of Philosophy* 79: 733–747.

Ginsberg, A. 1981. "Quantum Theory and Identity of Indiscernibles Revisited". *Philosophy of Science* 48: 487–491.

Heisenberg, W. 1926. "Schwankungserscheinungen und Quantenmechanik". *Z. Physik* 40: 501–506.

Luckenbach, S., ed. 1972. *Probabilities, Problems and Paradoxes*. Encino, Ca.: Dickenson.

Margenau, H. 1944. "The Exclusion Principle and its Philosophical Importance". *Philosophy of Science* 11: 187–208.

———. 1950. *The Nature of Physical Reality*, chapter 20. New York: McGraw-Hill.

Mittelstaedt, P. 1983. "Naming and Identity in Quantum Logic". Presented at the 7th International Congress Logic, Method. and Philosophy of Science, Salzburg, 1983.

Pauli, W. 1925. "Ueber die Zusammenhang des Abschlusses der Elektronengruppen im Atom mit der Komplexstruktur der Spektren". *Z. Physik* 31: 765–783.

Reichenbach, H. 1928. *The Philosophy of Space and Time*, section 43. English translation, New York: Dover, 1957.

———. 1956. *The Direction of Time*, section 26. Berkeley: University of California Press.

Salmon, W. 1969. Commentary on van Fraassen (1969). Summary in Luckenbach (1972), pp. 135–138.

Streater, R. F., and Wightman, A. S. 1964. *PCT, Spin and Statistics, and All That*. New York: Benjamin.

Sudarshan, E., and Mehra, J. 1970. "Classical Statistical Mechanics of Identical Particles and Quantum Effects". *International Journal of Theoretical Physics* 3: 245–251.

Teller, P. 1983. "Quantum Physics, the Identity of Indiscernibles, and Some Unanswered Questions". *Philosophy of Science* 50: 309–319.

van Fraassen, B. 1969. "Probabilities and the Problem of Individuation". Presented at the American Philosophical Association, 1969; published in Luckenbach (1972), pp. 121–135.

———. 1973. "Semantic Analysis of Quantum Logic", pp. 80–113 in C. A. Hooker, ed., *Contemporary Research in the Foundations and Philosophy of Quantum Theory*. Dordrecht: Reidel.

———. 1974. "The Einstein-Podoloski-Rosen Paradox". *Synthese* 29: 291–309.

———. 1981. "A Modal Interpretation of Quantum Mechanics", pp. 229–258 in E. Beltrametti and B. van Fraassen, eds., *Current Issues in Quantum Logic*. New York: Plenum.

———. 1982a. "The Charybdis of Realism: Epistemological Implications of Bell's Inequality". *Synthese* 5: 25–38.

———. 1982b. "Quantification as an Act of Mind". *Journal of Philosophical Logic* 11: 343–369.

Wigner, E. 1926/1927. "Ueber nicht kombinierende Terme in der neueren Quantentheorie". *Z. Physik* 40 (1926): 492–500; and (1927): 883–892.

Semantics and Quantum Logic

EDWARD MACKINNON

In an important survey of developments in the philosophy of science McMullin (1976) distinguished two major approaches to the interpretation of science. The first, the *logistic* approach, originally rested on two underlying assumptions. The first is a foundationalist one: a scientific theory must rest on a foundation of statements whose truth is established beyond reasonable doubt. The second, a logistic one, is that inferability is the proper test of a scientific claim. In modern times the first assumption has been modified in various ways. Thus, a hypothetical-deductive approach takes the foundational statements to be hypotheses subject to the test of confirmation and/or falsification. The second requirement, inferability, has been modified and expanded to accord a basic role to other forms of inference in addition to deduction from axioms.

These modifications concern methods of development, not the underlying ideal. This may be simply sketched. A scientific theory, particularly a physical theory, has a deductive structure. Modern logic supplies a rich assortment of tricks and tools for reconstructing such theories in a way that makes the inferential connections transparent. If different theories are logically reconstructed in a standard way, then one has a basis both for comparing different theories and for establishing general norms and criteria for appraising these theories. The practical conclusion following from these considerations is that a rationally reconstructed theory supplies the unit of appraisal which philosophers of science should use in treating such philosophical issues as truth, meaning, realism, confirmation, falsification, and other questions concerning theories as explanatory units. The general model of a rationally reconstructed theory has to be supplemented in different ways to adapt it to different functioning theories and different stages of development of a theory. These, however, are questions of detail that can be worked out with sufficient ingenuity and patience.

The chief shortcomings that McMullin finds with this approach concern the questions of fertility and coherence. The theories that admit of the clearest logical reconstruction are theories, like geometric optics or classical particle dynamics, that have ceased to grow and change. They may be fertile in the limited sense of explaining an already established body of facts; they are not *P-fertile,* in predicting new facts and laws. Fertile theories, ones that lead to new discoveries and increased understanding, are usually theories that are growing and changing. No fixed formulation can be taken as the canonical expression of such a theory.

Coherence is a notoriously slippery issue. Basically, it concerns the way a particular theory relates to the established body of scientific knowledge. If one thinks of each theory as an isolated system tested by its own observation reports, rather than by its relation to other theories, then it is difficult to give 'coherence' any clear meaning except in special cases of reducing one theory to another. Yet, it is also clear that such coherence considerations play a significant role in the process by which an on-going community of scientific inquirers appraises the acceptability of any new theory.

The second major approach to interpreting science, a *historicist* approach, has two rather different implementations. The first, a purely descriptive one, would trace the development of a theory from its inception through its various modifications. In such a purely historical context the question of which is the true or correct or real theory is no more meaningful than it would be for a paleobiologist to trace the evolution of the horse from Eohippus to Citation and try to decide which one is the real horse. Such questions are meaningful only if there are some sort of norms in terms of which one judges the goodness, adequacy, or whatever, of a theory.

The second implementation of the historicist approach is, not surprisingly, a normative one. Different normative patterns for the development of scientific theories have been proposed by Kuhn, Toulmin, Lakatos, and Laudan. These, as McMullin points out, have their own difficulties with the question of P-fertility. They also have extreme difficulty with the question of coherence, difficulties epitomized in the now notorious phrase, "the incommensurability of paradigms".

I agree with McMullin's conclusion that an adequate theory of scientific theories must treat the interconnection of three issues: the fertility of theories, especially predictive fertility; the unit of appraisal, an isolated set of timeless propositions, or a developing theory interpreted as part of scientific knowledge in general; and the relation between the realism that characterizes functioning science and the epistemological problems encountered by any doctrine of scientific realism. These problems have convinced most developmentalists and many reconstructionists that any doctrine of scientific realism is misguided.

Yet, even those who reject scientific realism as a *philosophical* position should supply some account of the role that scientific realism plays in functioning science. These are the issues that must be treated. But how?

I think that what is needed now is a sort of scissors action. One cuts down from above with the sort of general considerations just briefly sketched. One must also cut up from below through a detailed investigation of much more specific issues. In particular, one needs to consider aspects of individual scientific theories where both of the approaches mentioned can make a contribution. This supplies a concrete basis for working out the limits of validity of each approach. An exclusive top-down approach tends to select from science and its history only facts and issues congenial to the approach in question. This obviates any meaningful confrontation. In the remainder of this paper we will consider one such particular issue. After analyzing this in some detail we will return, at least briefly, to the more general issues in the background.

1. A Bohrian Semantics

The issue we wish to focus on concerns the meaning of some philosophically problematic terms in quantum mechanics and the way the problem of semantics is treated in this context. This problem will be treated from two perspectives. The first perspective, a historical one, will be treated quite briefly. The reason for the brevity is that what we are presenting here is a sort of summary skimmed off a more detailed development given elsewhere (MacKinnon, 1982a). The second perspective is more logistic, in McMullin's sense of the term. We will examine the way in which the same problematic concepts are treated in a logical reconstruction of quantum theory, or at least of a significant portion of quantum theory.

In a logistic approach to developing a metatheory of scientific theories one divides scientific terms into two classes, observational and theoretical. Many have pointed out that the distinction is not sharp. Shapere (1982b) has shown convincingly that the practices of the scientific community differ significantly from the presuppositions that philosophers bring to the interpretation of that practice. Yet one consequence of this distinction remains unchallenged. That is that the epistemologically problematic terms are the theoretical ones. The current debates on scientific realism center on the issue of whether crucial theoretical terms can be interpreted as referring to unobservable entities.

Bohr's epistemology, which should not be identified with the Copenhagen interpretation of quantum mechanics, grew out of his struggles to interpret the problematic concepts of functioning science.[1] With one apparent exception, which will be treated later, the epistemologically problematic concepts were *never* the theoretical terms. The problematic terms are classical terms, such as 'position' and 'momentum', which play an indispensable role in describing experiments and reporting results.

Classical terms used in classical contexts fit a normal distributive logic. So do the purely theoretical terms introduced into atomic theory and particle physics. Thus, a particle either has integral spin or half-integral spin. If a particle has a strangeness quantum number of 0, then it can not have a strangeness quantum number of 1. Such locutions fit the standard formulations of 'or' and 'if . . . then'. Suppose, however, that one is describing the trajectory of an electron in an ideal double-slit experiment (ideal, since there is no experimentally feasible way to perform this experiment for one electron). If the electron is recorded on a photographic plate then one can meaningfully say:

(1) Either the electron went through slit A or through slit B

and interpret (1) as one statement. It is not equivalent to the two statements:

(2) The electron went through slit A *or*
    The electron went through slit B.

Quantum logicians see the 'or' of (2) as the fulcrum supporting a better solution. Bohr thought that a solution of the epistemological problems indicated by (1) and (2) hinged on an analysis of the crucial concepts and the frameworks in which they function.

Bohr's epistemological development has been summarized elsewhere. The viable features of his position come from his protracted efforts to analyze the conditions of the possibility of unambiguous communication concerning atomic experiments. Now I will simply summarize the features of this analysis that have a bearing on the problem at issue. Since this is *not* intended as an historical account, the theses that follow will be labelled *MBS,* or minimal Bohrian semantics. It is a position derived from his work, but one which stands or falls on its merits.

1. *The philosophically problematic concepts are classical terms which play an indispensable role in describing atomic experiments and reporting results.* By "classical concept" we mean ordinary language terms like 'particle', 'wave', 'position', and 'momentum', which have been incorporated into classical physics.

2. *The meanings that these classical concepts have in an experimental context are determined by their usage in classical physics.* It is important not to confuse this with the standard distinction between theoretical and observational terms.

The pivotal question is: What are the conditions of the possibility of the un-ambiguous communication of information about atomic experiments? X performs an experiment and reports what she observed. A necessary, though not sufficient, condition for Y to understand her report is the sharing of a common language. In the mid-thirties Bohr was groping with problems of the nature and grounds of meaning in ordinary language, problems that came to the forefront of the Anglo-American tradition only in the mid-fifties. He never developed a theory of meaning in general. He focused on the concepts he found problematic, and then assumed that the method of clarification that worked for these would also work for less problematic concepts. However, the natural affinity between Bohr's epistemology and ordinary language analysis, particularly the work of Wittgenstein, Strawson, and Sellars, supplies a vehicle for extending and clarifying Bohr's epistemological position.

Two aspects of his position on this point require comment. First, he had totally abandoned any sort of referential theory of meaning. Thus, 'particle', a classical concept, could not get its meaning from any entities it referred to. Classical physics does not supply a basis for treating fundamental particles. Its meaning came from its usage in ordinary language and the idealization of that usage in classical physics. Secondly, unambiguous reference to absent or unobservable entities always presupposes the subject-object distinction implicit in ordinary language usage. When Bohr stressed this distinction he was not attempting to introduce subjectivity into quantum theory. Rather, he was explicating, in his own chiaroscuro style, that any claims of objectivity are grounded in a subject-object distinction.

3. *There is a complementarity between classical and quantum physics.* Semantically, quantum theory and its applications manifest an essential dependence on classical physics. All information about atomic systems and fundamental particles ultimately depends on performing experiments and reporting results. All observational reports are in classical terms, neglecting the quantum of action. Quantum theory cannot be understood as an isolated system. The reason for this is semantical: the unit of meaning is language as a whole.

With one apparent exception, the philosophically problematic concepts are classical concepts used in the quantum domain. These are the concepts Bohr wrestled with; these are the concepts that seem to necessitate a deviant logic. The one apparent exception concerns spin components. If the z-component of a particle's spin is known, then neither the x- nor the y-component can be known. 'Spin component' is not a classical concept. Yet, it requires the same sort of deviant logic that position and momentum attributions do.

Here a bit of history shows why this exception is more apparent than real. Bohr's epistemology is severely limited. It focused on the grounds of meaning of the concepts requisite for an unambiguous account of the results of a scientific experiment. These, Bohr insisted, must be classical concepts. Since 'spin' is not a classical concept and vanishes in the classical limit, Bohr concluded that spin cannot be directly measured. According to Rosenfeld, everyone except Pauli was shocked by this conclusion. Bohr's qualitative arguments were soon developed by N. F. Mott (1949, pp. 61–66) into a proof that it is impossible to measure the spin of an isolated particle or atom. Any measurement would require an interaction between the magnetic moment associated with spin and an electric field. However, within the limits of the indeterminacy principle the electrical interaction between the particle or atom and the field would blot out this weak magnetic interaction.

The resulting situation can be summarized as follows. On theoretical grounds, electrons within atoms were assigned spin values. Electrons, when detached from atoms, must also have spin values, even though they cannot be measured directly. A Stern-Gerlach experiment supplies an indirect basis for measuring the spin component of an atom (or its valence electron) in a particular direction. What one actually measures is angular momentum, a thoroughly classical quantity. Then, on theoretical grounds, one interprets this splitting of angular momentum components in an inhomogeneous magnetic field as differences in spin-component values of different atoms.

4. *There are further complementarities, especially between 'particle' and 'wave'.* Here again, any dependence on a referential theory of meaning renders this doctrine radically unintelligible. 'Particle' is at the center of a cluster of interrelated concepts. A particle travels in a trajectory, hits a target, impinges, is deflected, recoils, strikes a plate at a point. This is not because of what electrons are, but because of the way 'particle' functions in classical physics. A wave does not travel in a trajectory. It undulates, is reflected or refracted. It can interfere with other waves. But it cannot strike a plate at a point. Here again, we have a conceptual unpacking of a classical concept.

Sources of information about atomic systems can be roughly divided into two types. In the first, one hits a target with a projectile and then examines the debris. In the second, one examines the radiation that is produced, absorbed, or modified in atomic interactions. One of the two concept clusters, 'particle' or 'wave', is adequate—and indispensable—in describing any experiment and reporting results. Neither is adequate for handling all experiments. Which one is appropriate depends on the questions we put to nature.

5. *Each atomic experiment is an epistemologically irreducible unit.* If we revert back to the numbered sentences 1 and 2, we can illustrate this. If one

performs a double-slit experiment with both slits open, then one may affirm sentence 1 of an electron that hits the screen. The only way to determine which slit the electron went through is to change the experiment, to close one slit. Then one is not conceptually subdividing the original experiment; one is performing a different experiment.

Bohr eventually treated this as a purely semantical problem. Quantum theory rests on assumptions incompatible with classical physics. Yet, any observational reports require that an essential use be made of classical concepts and that these concepts preserve their classical meanings. The principle of complementarity guides the *extension* of classical concepts to quantum domains. The principle of epistemological irreducibility *restricts* their usage. To put this in jargon more familiar to philosophers, we may distinguish between pure language games, which involve only the relation of words to other words, and impure language games, which also include the physical context in which the language game functions. Then the principle of epistemological irreducibility becomes: the minimal language game in which classical concepts can be meaningfully used in observational reports about atomic systems is an impure language game involving both the system being studied and the apparatus through which it is studied. One can not interpret such reports as giving descriptive accounts of atomic systems detached from any means of observing them.

This summary suffers from two major limitations. First, it is just a summary, leaving out both the subtleties and shortcomings of Bohr's position as well as the justification for the theses presented. Second, and more significant, Bohr's semantics is radically underdeveloped. He went over and over the problematic concepts until he found a way of clarifying the grounds of meaning in a way that explained how the unambiguous communication of information about atomic systems is possible. He did not develop a general theory of meaning or attempt to relate his position to any theories of semantics or language.

Yet, such limitations notwithstanding, there is a clear contrast between a Bohrian semantics, or an explication of meaning in terms of use, and the two normative approaches considered earlier. The differences between Bohrian semantics and the theories of semantics characterizing a normative historicist approach is clearest in the case of Kuhn's earlier position (1962). Kuhn pictured the development of science in terms of a series of conceptual revolutions and focused on semantic problems concerning terms whose meaning in one theory seems incommensurable with their meaning in another theory. Bohr also interpreted the development of scientific theories in terms of conceptual revolutions. What he found philosophically problematic, however, was not the conceptual revolution itself but the consequent reorganization of physical concepts necessary to achieve a coherent integration of the new and the old. This led to a focus on a semantic clarification of the concepts which are common to the new and old theories, rather than on concepts that differ in the two theories. This is more a difference in emphasis than a clash in doctrine. There is no significant clash between Bohrian semantics and Kuhn's revised epistemology.

The clash between Bohrian semantics and the semantics that characterizes logistic interpretations of scientific theories is central rather than peripheral. Bohr did not take the mathematical structure to be the core of a theory. The mathematical structure is considered an algorithm. The crucial points requiring clarification are the concepts used to describe physical reality and to report experimental results. This clarification has two prongs: a lower one, showing how the concepts used to report observational results are embedded in ordinary language; and an upper prong, showing how these concepts can be extended to and restricted in new contexts. This focus on ordinary language and the grounds of meaning puts Bohrian semantics very much in touch with the main thrust of Anglo-American philosophy. However, it clearly puts it out of touch with the logistic methods that see a mathematico-logical structure as the core of a physical theory and treat semantic issues by imposing a formal semantics on a theory reconstructed as a formal system. The remainder of this article will be concerned with the leading attempt to treat the problematic concepts of quantum theory through such a reconstruction.

2. Quantum Logic

The tradition loosely labeled "quantum logic" also treats the problem of the proper interpretation of quantum mechanics. It does so, however, from a radically different perspective. The focus of attention is on a formal reconstruction of a fragment of nonrelativistic quantum mechanics, rather than on physics as practiced by quantum physicists. Quantum logic grew out of the effort of Birkhoff and von Neumann (1936) to interpret the lattice of subspaces of a Hilbert space as a logic. It was revived by Mackey (1963) and transformed, in the sense that the logic was interpreted as a logic of sentences with Hilbert space structures supplying one model for the logic. The brief summary that follows presents the core of the standard development while skipping some disputed issues not germane to our present considerations.

We assume a language, $\mathfrak{L}$, with predicates, P_m^n, and the logical connectives, $(\sim, \&, \vee, \supset, V)$. A system of sentences based on these connectives and their standard interpretations does not have a Hilbert space model which fits quantum theory. So, it is necessary to impose a new structure on $\mathfrak{L}$, or on a proper subset of $\mathfrak{L}$. A standard structure is an axiom system S. S is an axiom system if and only if (iff) $S = <\mathcal{A}, \mathcal{R}>$, where $\mathcal{A}$ is a set of well-formed formulas (wffs) of S and $\mathcal{R}$ a set of inference rules. I assume that this is a familiar notion.

It is also possible to have the weaker notion of a logic. A logic is a set of wffs closed under certain rules of inference. We will let $A \rightarrow B$ stand for "B may be inferred from A". Now, following Dalla Chiara (1977), we may set up axiomatic schemata for minimal quantum logic (henceforth MQL). We first define the set of all formulas with an implicational basis (henceforth, the I-formulas of $\mathfrak{L}$).

Definitions

The set of I-formulas of $\mathfrak{L}$ is the smallest set, Y, of formulas such that:
(a) for any A and $B \in Y$, $(A \rightarrow B) \in Y$;
(b) for any $A, B \in Y$, $\sim A$, $\sim B$, $A \& B$, $(Vx) A \in Y$.
An I-formula will be denoted by $\underline{A}$.
Now we present Dalla Chiara's systematization of MQL.

Logical Axioms of the MQL Calculus

(1) $A \rightarrow A$
(2) $(A \rightarrow B) \rightarrow ((B \rightarrow C) \rightarrow (A \rightarrow C))$
(3) $A \& B \rightarrow A$; $A \& B \rightarrow B$
(4) $(C \rightarrow A) \rightarrow ((C \rightarrow B) \rightarrow (C \rightarrow A \& B))$
(5) $A \& \sim A \rightarrow B$
(6) $\sim\sim A \rightarrow A$; $A \rightarrow \sim\sim A$
(7) $(A \rightarrow B) \rightarrow (\sim B \rightarrow \sim A)$
(8) $(A \rightarrow (B \rightarrow C)) \rightarrow ((A \rightarrow B) \rightarrow (A \rightarrow C))$
(9) $(Vx)A(x) \rightarrow A(x/t)$
(10) $\underline{A} \rightarrow (B \rightarrow \underline{A})$
(11) $(A \rightarrow B) \rightarrow (B \rightarrow (A \vee (\sim A \& B)))$

Rules for the MQL calculus

(i) $\dfrac{A,\ A \rightarrow B}{B}$

(ii) $\dfrac{A}{B \rightarrow A}$

(iii) $\dfrac{A \rightarrow B(x)}{A \rightarrow (Vx)B(x)}$

Here we make the standard assumptions concerning suppression of parentheses and *t*'s being free for *x* in *A*. The first six axioms are essentially syntactical definitions of the logical connectives and play the same role as the familiar introduction and elimination rules. Axioms (7), (8) and (9) are schemata used in a Hilbert-style axiomatization of the first-order predicate calculus. Axiom 10 replaces the familiar classical principle that a true statement is implied by any sentence by restricting it to the case where $\underline{A}$ is an I-formula. This is a weakening of standard logic. Axioms (1) through (10) constitute an *orthologic*. (See Goldblatt, 1974.) Rule (ii) is introduced to compensate for this limitation.

Axiom (11) restricts this orthologic to an orthomodular logic. To see its significance, consider the standard rule for ∨-elimination. If a proof contains $(A \lor B)$ and $A \rightarrow C$ and $B \rightarrow C$, then one may eliminate $(A \lor B)$ in favor of *C*. This presupposes, to use semantic terms, that if $(A \lor B)$ is true, then either *A* is true or *B* is true or both are true. This is precisely the assumption we wish to avoid. It is replaced by the weaker form (11), which does not make '∨' distributive.

3. Informal Semantics

We have not yet given this MQL any interpretation which would allow us to treat issues of meaning and truth. In accord with the usual methods of logical reconstruction of a scientific theory we might take MBS as the intended interpretant of MQL. This assumption, however, quickly breaks down. The juxtaposition of MBS and MQL seems to present a formidable obstacle to any such attempt. One of the crucial points of MBS is that quantum assertions are highly context dependent, where the context is an experiment, whether actual or *gedanken,* including both the system being studied and the apparatus used to study it. A statement that is meaningful in one context, such as an ascription of position, would be meaningless in a context where momentum is being measured.

A logical system is one that is closed under logical connectives. This presupposes that it is possible to relate any two sentences, A and B, by such logical connectives as '&' or '$\rightarrow$' to get sentences like 'A & B', '$\sim A \rightarrow \sim B$', etc. This is purely syntactic. However, if the intended interpretation of MQL includes or even simply relates in a significant way to MBS, then there is a definite clash. This is not simply a logical possibility; it is a historical fact. Bohr, Heisenberg, and Pauli, the framers of the Copenhagen interpretation, had no use for logical reconstructions of quantum theory. Quantum reconstructionists refer to Bohr's semantics as confused, puzzling, obscure, or perverse — or they simply ignore it.

Here we are more concerned with building bridges than with creating obstacles. To interrelate MBS with attempted reconstructions we will treat semantics on three separate levels: informal semantics, formal semantics, and the informal interpretation of the formal semantics. By 'informal semantics' we mean semantic notions of the pre-logical language, or the semantics that is presupposed prior to the imposition of a formal structure. In discussing informal semantics we are primarily concerned with the obstacle presented by the clash between MBS and any logical reconstruction. MBS is interpreted as a clarification of the meaning of crucial terms used in observation reports about atomic systems. This leads to a doctrine of meaning through use and of ordinary language as the ultimate ground of meaning. For this reason its relation to a logical reconstruction should be seen as an extension of the way ordinary language usage relates to logical reconstruction.

The meaning that such terms as 'and', 'or', and 'all' have in English is determined by their use in English, not by the definitions of the logical connectives. The fact that '$\vee$' and '$\supset$' only pick up one of a family of meanings associated with 'or' and 'if . . . then' is familiar. Something similar obtains for the other logical connectives. The sentence "He got sick and died" is not equivalent to "He died and got sick", though "p & q" is equivalent to "q & p". As Zeno Vendler (1967, pp. 70–96) has pointed out, 'all', 'each', 'every', 'any', 'the', and 'a' have subtle but significant differences in meaning. Yet, in a logical reconstruction, all can get mapped into $\forall x$.

The general situation may be summarized by saying that a logical reformulation achieves syntactical precision at the cost of semantical impoverishment. The crucial question now is: what semantical impoverishment of MBS tailors it to MQL? A comparison of MBS and the tradition of quantum logic suggests one answer. A basic feature of MBS is the doctrine that the meaning of an observation report is dependent on the context of an experiment considered as an epistemologically irreducible unit. To fit this into a logical reconstruction it should be replaced by the weaker notion of *compatibility*. Since the significance of this notion is disputed I will clarify the meaning I am attaching to it.

Let's begin with the physics. Given a system to be studied, two quantities characterizing the system are compatible if the measurement of one does not preclude the measurement of the other, or if the order in which they are measured does not affect the results of measurement. This is symbolized '$A \downarrow B$'. This physical idea is carried over to the mathematical formalism of quantum mechanics in the doctrine, or definition, that two quantities are compatible if the operators representing them commute. In spite of this reference to the standard mathematical formalism, 'compatible' is still being treated as a prelogical notion. It is a notion that is part of functioning quantum mechanics prior to the imposition of any logical reconstruction. This should be distinguished from the prelogical notion of 'orthogonal'. Thus the states of an atomic system are orthogonal in that the attribution of one state excludes the attribution of other states. Here we prescind from complications associated with the superposition principle. Yet, statements attributing orthogonal values can be compatible. It is quite meaningful to say, for example, that a hydrogen atom is either in the ground state or in the first excited state.

Suppose that two sentences, A and B, are incompatible. What significance should be attached to the conjunction, 'A & B'? MBS would claim that the conjunction is not physically meaningful. A context in which A can be measured precludes any measurement of B. However, banishing such conjunctions vitiates the idea of setting up a logical system. One might claim that the conjunction is always false. This does not fit well with either the physics or the logic since it seems to attribute a definite truth value to a statement that has no physical meaning.

Compatibility is not a property of a sentence, but of its relation to other sentences. We can associate with any sentence two classes: a coassertion class and a codenial class. If A and B are incompatible, then it never makes sense to assert (A and B). However, if A is known to be false, it makes sense to deny (A and B). Similarly, if A is known to be true it makes sense to assert (A or B). However, a denial of (A or B) only makes sense when $A \downarrow B$.[2]

4. Formal Semantics

Discussions of the semantics of quantum logic involve three major problems. The first is setting up a model which allows an interpretation of the wffs of MQL and an assignment of truth values. The second is the question of whether the logical connectives used in quantum logic have the same meaning as they have in classical logic. The third is a semantics for a modal quantum logic. We are concerned only with the first issue.

A model for MQL is under a nonlogical constraint. It must fit the mathematical structures used in quantum theory. The procedure usually followed is to construct a mathematical model and then admit that its relation to functioning quantum mechanics is problematic. For this reason we will begin with the problems before constructing the model.

Loosely speaking, the basic building blocks in MQL are atomic sentences. The intended interpretation is that an atomic sentence, A, represents an elementary observation of some quantity Q, characterizing a system. Thus A could be represented by (q,e), where q is a numerical value attached to Q and e is a Borel set (the least set of all open and closed intervals on the real line which is closed under set-theoretic complements and countable unions). A is true if q is in e, false otherwise.

With the same degree of laxity, we may say that the basic building blocks in our mathematical system are vectors in a Hilbert space and linear manifolds constructed from these vectors. A vector represents a pure state of a system. If we could simply map atomic sentences in MQL into vectors in the Hilbert space representing a system, we would have a nice neat representation.

This neat correspondence is untenable on physical grounds. Any measurement of the state of a system requires an interaction with a measuring apparatus. No interacting system is in a pure state. Further, a pure state is not in general characterized by the specification of one quantity, Q, but by all the quantities which can be simultaneously specified. This leads to the idea of a complementary relation between observation and state specification. A pure state is, in principle, unobservable.

There are, in general, three ways around this difficulty. The first is to develop a systematic reconstruction of quantum theory (not just quantum logic) which does not accord 'state' a foundational role, but just works with quantities that are observable, at least in principle. Then 'state of a system' may be introduced as a derived or defined concept based on an idealized extrapolation from 'quantity'. This has been developed in slightly different ways by Scheibe (1973) and by van Aken (forthcoming).

A second approach is to give an operational interpretation of MQL. This has been developed in different ways by Jauch (1968), Mittelstaedt, and Stachow (Mittelstaedt, 1974, 1976). In addition to the well-known difficulties of operationalism, it seems to me that this approach determines axioms on the basis of observations only by slighting the essential role that idealizations and extrapolations play in any attempt to go from an observational basis to an axiomatics of quantum mechanics. We will summarize the ones that play a role in the interpretation of quantum logic.

A quantity, Q, characterizing a physical system is represented by a linear operator, which must be Hermitian and self-adjoint for its expectation value to be a real number. If we begin with bounded systems, then we can find quadratically integrable functions, ψ, such that

$$Q\psi = q\psi,$$

where Q is used both for a quantity and the operator that represents it, and q is a number, an eigenvalue. This can be extended to non-bounded cases and continuous, rather than discrete, spectra by various mathematical tricks. In both cases ψ can be taken as representing states of a system for which Q has definite values. This process leads to the idealization: Every physical quantity is represented by a Hermitian operator with a complete set of observables. This can be extrapolated into a definition of quantities in terms of operators: every self-adjoint Hermitian operator corresponds to an observable quantity. The eigenvalues and eigenfunctions characterizing the system can be represented by an expansion, $\psi = \sum_{i=1}^{\infty} c_i \varphi_{ij}$ where $\int \varphi_i^* \varphi_j d\tau = \delta_{ij}$ and $c_i = \int \varphi_i^* \psi d\tau$.

The most convenient way to express these wave functions is through Dirac notation in Hilbert space,

$$| \psi > = \xi_i | \xi_i >.$$

Here $| \xi_i >$ is a vector, but not in general a basis vector. It is projected out by the projection operator,

$$P_i = | \xi_i > < \xi_i |,$$

with the properties

$$\sum_{i=1}^{\infty} P_i = 1, \qquad\qquad P_i^2 = P_i \text{ (idempotent)}.$$

To relate this to a feasible experiment we will consider the Stern-Gerlach experiment mentioned earlier. A hydrogen atom passing through an inhomogeneous magnetic field can be represented by the sum $\psi = U_+(r) + U_-(r)$, where

$$U_+(r) = V(\mathbf{r})S, \quad U_-(r) = V'(\mathbf{r})S,$$
$$V(\mathbf{r}) = \sum_{n,l} R_{n,l}(r) Y_l^m(\theta,\varphi)$$
$$V'(\mathbf{r}) = \sum_{n,l} R'_{n,l}(r) Y_l^m(\theta,\varphi).$$

Here we have used S for the spin component, the only aspect of the wave function we need. This can be expressed in terms of four projection operators:

$$S_x = \hbar/2 \ (Q_1 - Q_2) \qquad S_y = i\hbar/2 \ (Q_1 + Q_2)$$
$$S_z = \hbar/2 \ (P_1 - P_2),$$

where

$$Q_1 = \tfrac{1}{2}\begin{pmatrix} 1 & 1 \\ 1 & 1 \end{pmatrix}, \qquad P_1 = \begin{pmatrix} 1 & 0 \\ 0 & 0 \end{pmatrix}$$
$$Q_2 = \tfrac{1}{2}\begin{pmatrix} 1 & -1 \\ -1 & 1 \end{pmatrix} \qquad P_2 = \begin{pmatrix} 0 & 0 \\ 0 & 1 \end{pmatrix}.$$

Each of these is an idempotent operator. Thus P_1 projects onto the subspace $\begin{pmatrix} 1 \\ 0 \end{pmatrix}$. It correlates with the theoretical ascription of spin up to the atom's electron (for a hydrogen atom in the ground state) and with a measurement of a positive z-component for the hydrogen atom's total angular momentum. We will abbreviate this by saying 'spin$_z$ up'.

Now suppose we isolate the part of the incoming beam of hydrogen atoms that go through the Stern-Gerlach experiment and manifest spin$_z$ up. By another suitably oriented Stern-Gerlach magnet we may measure the x-component of these atoms, which is projected by the operator, Q_1. It is tempting to interpret the positive component of the resulting beam as being made up of atoms with spin$_z$ up and spin$_x$ up.

As McMullin might say, lead us not into temptation. This particular temptation is overcome by both theory and experiment. Theory tells us that P_1 and Q_1 are incompatible, because they do not commute:

$$P_1 Q_1 = (\tfrac{1}{2})\begin{pmatrix} 1 & 1 \\ 0 & 0 \end{pmatrix}, \qquad Q_1 P_1 = (\tfrac{1}{2})\begin{pmatrix} 1 & 0 \\ 1 & 0 \end{pmatrix}.$$

Experiments indicate that if we were to measure spin$_z$ for atoms with spin$_x$ up, half of them would have spin$_z$ up and half would have spin$_z$ down. Accordingly, following Jauch, we would define the meet of P_1 and Q_1, corresponding to the proposition that the hydrogen atom has spin$_z$ up and spin$_x$ up, as

$$P_1 \cap Q_1 = \lim_{n \to \infty} (P_1\, Q_1)^n$$
$$= \lim_{n \to \infty} (\tfrac{1}{2})^n \begin{pmatrix} 1 & 1 \\ 0 & 0 \end{pmatrix} = 0.$$

Physically this would correspond to an infinite series of Stern-Gerlach apparatuses which alternately select spin$_z$ up, spin$_x$ up, spin$_z$ up, etc. This is a clear example of an idealized extrapolation from the process of measurement.

After these preliminary considerations, we may now define the lattice that serves as a model for interpreting MQL. The basic elements are projection operators which stand in a one-to-one correspondence with subspaces of a complete, separable, complex vector Hilbert space. We define the meet of any two elements as

$$P \cap Q = \lim (P\, Q)^n.$$

Then the meet of any two incompatible elements is 0 and, because of idempotency, $\lim (P\, P)^n = P$. We define orthogonality in terms of the relationship between subspaces (and use the subspace-projector isomorphism to get orthogonality for projectors): $Y^\perp = X$, if X is the largest subspace such that every vector in X is orthogonal to every vector in Y. Two vectors are orthogonal if their inner product vanishes. For the join of two subspaces, $P \cup Q = (P^\perp \cap Q^\perp)^\perp$. The final primitive notion, $\leq$, is defined in terms of partial inclusion of subspaces and leads to the corresponding relation for projectors, $P_x \leq P_y$ if X and Y are subspaces and $X \leq Y$. Compatibility is taken as a pre-logical notion. If $P \overset{.}{\flat} Q$ within a model, then within that model

$$(P \cap Q^\perp) \cup P = (Q \cap P^\perp) \cup P.$$

Let S be a set of projection operators, $\mathcal{B}$ be the algebra of projection operators partially ordered by the inclusion relation $\leq$. Then for any $P, Q \in S$:

1. $P \cap Q \leq P$
2. $P \cap Q \leq Q$
3. If $P \leq Q$ and $P \leq R$, then $P \leq Q \cap R$
4. $P \leq P \cup Q$
5. $Q \leq P \cup Q$
6. If $P \leq Q$ and $R \leq Q$, then $P \cup R \leq Q$.

So this partially ordered set of projectors is a lattice. This lattice is ortho-complemented since

7. $P \cup P^\perp = 1$ (1 = the Hilbert space)
8. $P \cap P = 0$ (0 = the null space)
9. $P = P^{\perp\perp}$
10. If $P \leq Q$, then $Q^\perp \leq P^\perp$.

Finally, the lattice is orthomodular since

11. If $P \leq Q$, then $Q = P \cup (P^\perp \cap Q)$.

To develop the formal, or algebraic, semantics we postulate an ordered quadruple, $< \mathrm{FS}, \mathcal{B}, h, \vee >$, where $\mathcal{B}$ is an ortholattice of projection operators. FS is a formal system, $\mathrm{FS} = < \mathcal{L}', C, A >$, where $\mathcal{L}'$ is $\mathcal{L}$ expanded to include individual constants, c, such that $h(c) = d$, for all $d \in H$. h is a function mapping elements of L into H such that $h(c) = d$, and $h(P^n)$ is an n-ary relation of the elements, d, of H.

For any sentence, A, of FS, we require that: $h(\sim A) = h(A)^\perp$; $h(A \ \& \ B) = h(A) \cap h(B)$; $h((\forall x)A) = \cup \{h(A(x/d))\}$, where x/d indicates the substitution of d for x. $\vee$ is a valuation proper to a model. On physical grounds we require that A is true in a model if A assigns a value q to Q and $h(A) = P_Q$, where P_Q projects a subspace for which $Q\psi = q\psi$. If $h(A)$ makes A true in a model, $\mathcal{M}$, it also makes $h(A \rightarrow B)$ true iff $h(A) \leq h(B)$.

This definition of 'true' is unambiguous when applied within a maximum Boolean subalgebra, which represents a maximal set of compatible properties. Can it be extended beyond this? Since this is basically a question of the physical significance to be accorded such an extension, we will consider the two clearly contrasting positions. In the "ignorance" interpretation, every property of a system objectively has a value, though we are ignorant of the values incompatible with those proper to a particular experimental set-up. This is at the heart of the Einstein-Podolsky-Rosen paradox and of most hidden variable theories. In addition to the difficulties besetting hidden variable theories, there is a purely logical difficulty. The difficulty, stemming from Gleason's theorem (Gleason, 1957) and the Kochen-Specker theorem (Kochen, 1967), was recently given an elegant formulation by Pitowsky (1982). Any attempt to have an unrestricted truth function defined on a non-distributive ortholattice generates a contradiction.

In the Copenhagen interpretation any specification of properties is highly context dependent. More technically, the only properties that are relatively context independent are those represented by operators which commute with the Hamiltonian. When we have a complete specification of a state, then we have a specification of all the properties that are mutually compatible. Any further specification of properties in this context is meaningless. Meaningless statements cannot be assigned truth values. We will, accordingly, limit the simultaneous assignment of truth values to compatible properties, or to the maximum Boolean subalgebra representing them.

The notion of truth may be extended to semantic entailment and to validity. A valid formula is one that is true in every interpretation of every model. For sets of formulas, Γ and Δ, we say that Γ semantically entails Δ with respect to a model $\Gamma \models_{\mathcal{M}} \Delta$ if any assignment of values that makes Γ true also makes all the sentences of Δ true. Using these notions it is possible to show that quantum logic is[3]

$\quad$ *sound*: $\quad$ if $\Gamma \vdash_{MQL} \Delta$, then $\Gamma \models_{\mathcal{M}} \Delta$,

and *complete*: $\quad$ if $\Gamma \models_{\mathcal{M}} \Delta$, then $\Gamma \vdash_{MQL} \Delta$.

5. Informal Interpretation of the Formal Semantics

This algebraic semantics yields, in van Fraassen's terms, a partial interpretation of FS. It is essentially a mapping of FS onto the ordered octuplet, $< \mathcal{B}, \leq, \cap, \cup, \perp, 1, 0, \flat >$. This still leaves the question of how FS relates to physical reality. This is the basic question treated in the informal interpretation of the formal semantics. I will try to clarify the underlying issues by contrasting two different strategies used to answer this question.

The more or less standard strategy is to assume that, at least for philosophical purposes, a logical reformulation of a theory may be considered a revised version of the theory in question. Most reconstructionists would readily admit that a logical reconstruction is an unwieldy instrument for solving problems *within* physics or for interpreting experiments. The contention is that when one is asking questions *about,* rather than *within,* a theory, questions concerning interpretation, truth, physical significance, relation to other theories, and similar questions, then the logically reconstructed theory supplies the clearest basis for unambiguous answers. Often this contention is not explicit, but it is implicit in the way logical reconstructions are used. The whole point and purpose of a logical reconstruction, many would argue, is to supply a basis for answering just such questions.

This mode of interpretation leads by simple, straightforward steps to a type of first-order referential realism. If, for philosophical purposes, the logically reconstructed theory can be considered a revised version of the theory being considered, then the theory should be interpreted the way any formal system is interpreted. One defines a morphology by postulating a domain of objects so that constants in the formal system stand for individuals, n-place predicates stand for n-place relations among these individuals, and parameters stand for arbitrary individuals. (See Thomason, 1970, ch. 11.) A valuation proper to this morphology serves to establish the truth values of sentences.

In the present case this would mean that the formal system, FS, is now interpreted through two models: a mathematical model constructed from Hilbert-space projection operators and a physical model, the physical system which this Hilbert space represents. There is, however, a widely shared agreement that this sort of referential realism does not work in the present case. The conclusion generally drawn is that the difficulty lies, not with the mode of interpretation, but with the doctrine of realism. As van Fraassen has forcefully argued (1980) and Putnam has reluctantly concluded (1977), there is no consistent way to defend a realistic interpretation of quantum logic. This leads to a mode of interpretation which could be called 'split-level operationalism'. The functional realism of ordinary language is admitted for familiar middle-size objects, and the functional realism of classical physics is accepted for the objects treated by classical physics. However, one gives an operational interpretation for the domain of objects proper to quantum theory.

In my opinion the underlying difficulty here lies in the mode of interpretation. It is misguided in general and impossible in the present case. Since, for present purposes, I only need the second point, I will merely indicate, in summary form, the reasons why I think it misguided to interpret a scientific theory by taking a logical reconstruction as basic and applying standard procedures for interpreting formal systems. While doing this I will indicate the special significance such criticisms may have for quantum theory.

First this mode of interpretation inevitably involves some sort of referential semantics. The basic word-world connection is between terms in a formal system and the things they denote. Truth builds on denotation, since any valuation presupposes a morphology. Donald Davidson has forcefully argued that one should reverse this procedure in interpreting ordinary language. For semantic purposes truth should be accepted as basic and reference accorded a derivative status (Davidson, 1977). This accords well with Sellars's arguments against any denotational semantics (Sellars, 1979). Elsewhere it has been argued that the semantic practices implicit in functioning physics accord truth, rather than reference, a basic role (MacKinnon, 1982b; Shapere, 1982a).

This is the consideration that renders this mode of interpretation impossible in the present case. The concepts which necessitate a deviant logic are *classical concepts.* These are concepts whose meanings are fixed by ordinary language usage and the idealization of this usage proper to classical physics. Here, it is essential that meaning be determined by usage, not reference. Thus, 'wave' and 'particle' form the cores of complementary clusters of concepts used to describe and report atomic experiments. Neither can be considered as referring, for classical physics does not treat real atoms or fundamental particles. When these terms are used in quantum contexts, their meaning is already set. They can, of course, be used to refer. But, such referential usage does not ground meaning; it presupposes that meanings are already established.

Second, this mode of interpretation gives a misleading account of the way scientific theories actually function. Interpretation is viewed exclusively as a relation between a theory and the domain of objects through which it is interpreted. In practice, however, the interpretation of a theory depends in a crucial way on its relation to a network of theories. In the present case, for example, a complementary relationship between classical and quantum physics plays an indispensable role in interpreting quantum physics.

Third, an exclusive reliance on this mode of interpretation almost invariably induces an epistemological naiveté. First-order referential realism treats interpretation as a relation between terms in a formal system and things in themselves. This naiveté is obscured by a retreat to split-level operationalism. In the Copenhagen interpretation, what a theory systematizes is what can be *said* about objects in a meaningful and, we hope, true fashion. Even when interpreting theories, we are still suspended in language.

The defense of this mode of interpretation tends, in practice, to be a rather occupational argument. This is what logicians do. This is reenforced by the rhetorical question: How else can a formal system be interpreted? One who disagrees with this mode of interpretation should reasonably be held to present a meaningful alternative. This I will attempt to do, albeit in a rather sketchy fashion.

I think that the best approach to answering this comes from a consideration of what is actually done in a logical reconstruction. A functioning scientific theory has as its object of study some part or aspect of physical reality. In a rational reconstruction the object being studied is not physical reality; it is a scientific theory. The theory becomes an object of study by being detached from its natural environment, its place in a loosely interrelated network of theories, and by being considered as an isolated unit.

Vico claimed that the mind of man can fully understand only what the mind of man has itself created. Something like this goes on at every level of theory construction. We try to understand aspects of physical reality by reconstructing them in a scientific theory. We try to understand the theory as an object, rather than a tool, by reconstructing the theory. A logical reconstruction can prune the luxuriant ontology of functioning theories and clarify the minimal basis the theory requires in its axioms and ontic commitments. It supplies the best basis for asking and answering questions concerning consistency, completeness, and independence of axioms.

A logical reconstruction, however, does not supply a basis for asking and answering questions concerning the relation of a theory to the objects treated by the theory. A logical reconstruction proceeds by detaching a theory from its normal functioning and from its relation to the objects it is a theory of. A philosopher interested in treating the way a scientific theory relates to the objects it is a theory of must focus on the functioning scientific theory, not on its logical reconstruction. A rationally reconstructed theory does not supply an adequate basis for any treatment of the issues of meaning, truth, or the ontological implications of a theory.

This brings us back to the general considerations McMullin has raised concerning the implicit realism of functioning science, the coherence of scientific knowledge, and the predictive fertility of scientific theories. The primacy which I am according functioning, rather than reconstructed, science in any consideration of the semantics of scientific knowledge supports a sort of scientific realism. It is a minimal and purely analytic form of realism. The term 'real' is an ordinary language term. Its meaning is determined by its usage in ordinary language. Thus, horses are real but centaurs are not; experiences

are real but illusions are not. When physicists use the term 'real' they are relying on established usage to set the meaning of the term, not on theories of
reality. Thus, physicists, not caught up in philosophical reflections, would normally say that electrons are real, quarks are probably real, Higgs bosons are
uncertain, and tachyons are probably not real.

Such an analytic clarification of the way 'real' and related terms function
does not, of course, suffice to answer the questions philosophers raise concerning objective reality, or things in themselves. It does, however, supply constraints. It constrains one to consider scientific knowledge, even knowledge
of the "theoretical entities" treated in atomic physics, as a part of human knowledge in general, not as a separate compartment handled by any sort of split-
level operationalism.

These considerations do not settle, or even properly treat, the issues of
coherence and predictive fertility. They do, however, help remove the chief
obstacle blocking any adequate consideration of the coherence of scientific knowledge. This is an image of scientific knowledge as a collection of formal systems
each interpreted through its own proper domain along the lines that formal
systems are treated in logic. With this scientific image banished it may be
possible to develop a more adequate image in which a central role is played
by ordinary language together with its various extensions, refinements, and
restrictions; while formal reconstructions of scientific theories are accorded
a significant but derivative status. Let us hope that McMullin's further work
will lead to the elaboration of such a coherent account of scientific knowledge.

Notes

1. The Copenhagen interpretation grew out of discussions between Bohr, Heisenberg,
and Pauli in Copenhagen in 1926–27. This "orthodox" interpretation, summarized in and dis-
seminated through Pauli's 1933 *Handbuch* article, includes: the uncertainty principle; the idea
that photons, electrons, and other particles exhibit both wave and particle properties; the probabilistic interpretation of the wave function; the correspondence between eigenvalues, given
by the theory, and the results of measurement; the complementary relationship between the
Heisenberg and Schroedinger representations; and the correspondence principle conclusion that
the results of quantum mechanics merge with classical mechanics in the limits of large quantum
numbers. These ideas are now a familiar part of quantum theory. In 1927 most of them seemed
novel and a bit bizarre. Bohr's epistemological position on the primacy of language, the grounds
of meaning, the subject-object distinction, and related issues should not be considered part
of the Copenhagen interpretation. Then one may accept the Copenhagen interpretation while
rejecting Bohr's epistemology.

2. My treatment of compatibility adapts ideas developed by Hacking (1980) and Cherna-vaska (1982).

3. The soundness and completeness of an orthologic was established by Goldblatt (1974). Further proofs were given by Dalla Chiara (1976) and Hacking (1980).

References

Adler, C. G., and Wirth, J. F. 1983. "Quantum Logic". *American Journal of Physics* 51: 412–417.

Birkhoff, G., and von Neumann, J. 1936. "The Logic of Quantum Mechanics". *Annals of Mathematics* 37: 823–843.

Bub, J. 1974. *The Interpretation of Quantum Mechanics*. Dordrecht: D. Reidel.

Chernavaska, A. 1982. "How to Get a Logical Structure Out of a Physical Theory". Unpublished.

d'Espagnat, B. 1976. *Conceptual Foundations of Quantum Mechanics*. 2nd rev. ed. New York: Benjamin.

Dalla Chiara, M. 1976. "A General Approach to Non-Distributive Logics". *Studia Logica* 35: 139–162.

———. 1977. "Logical Self Reference, Set Theoretical Paradoxes and the Measurement Problem in Quantum Mechanics". *Journal of Philosophical Logic* 6: 331–347.

Dalla Chiara, M. L., and Metelli, P. A. 1982. "Philosophy of Quantum Mechanics", in Guttorm Floistad, ed., *Contemporary Philosophy: A New Survey*. Vol. 2: *Philosophy of Science*, pp. 219–247. The Hague: Martinus Nijhoff.

Davidson, D. 1977. "Reality without Reference". *Dialectica* 31: 248–258.

Friedman, M., and Glymour, C. 1972. "If Quanta Had Logic". *Journal of Philosophical Logic* 1: 16–28.

Gleason, A. M. 1957. "Measures on the Closed Subspaces of a Hilbert Space". *Journal of Mathematics and Mechanics* 6: 885–893.

Goldblatt, R. I. 1974. "Semantic Analysis of Orthologic". *Journal of Philosophical Logic* 3: 19–35.

Hacking, I. 1980. "Why Orthomodular Quantum Logic is Logic". Unpublished.

Jauch, J. 1968. *Foundations of Quantum Mechanics*. Reading, Mass.: Addison-Wesley.

Kochen, S., and Specker, E. P. 1967. "The Problem of Hidden Variables in Quantum Mechanics". *Journal of Mathematics and Mechanics* 17: 59–67.

Kuhn, T. S. 1962. *The Structure of Scientific Revolutions*. Chicago: University of Chicago Press.

Mackey, J. 1963. *The Mathematical Foundations of Quantum Mechanics*. New York: Benjamin.

MacKinnon, E. 1979. "Scientific Realism: The New Debates". *Philosophy of Science* 46: 501–532.

———. 1981a. "The Interpretation of Quantum Mechanics: A Critical Review". *Philosophia* 10: 89–124.

———. 1981b. "Niels Bohr on the Unity of Science", in *PSA 1980* 2: 224–246. East Lansing, Mich.: The Philosophy of Science Association.

———. 1982a. *Scientific Explanation and Atomic Physics*. Chicago: University of Chicago Press.

———. 1982b. "The Truth of Scientific Claims". *Philosophy of Science* 49: 437–462.

McMullin, E. 1970. "The History and Philosophy of Science: a Taxonomy of Their Relations", in R. Stuewer, ed., *Historical and Philosophical Perspectives of Science*, pp. 12–67. Minneapolis: University of Minnesota Press.

————. 1974a. "Empiricism at Sea", in *Boston Studies in the Philosophy of Science* 14: 21–32. Dordrecht: D. Reidel.

————. 1974b. "Logicality and Rationality: A Comment on Toulmin's Philosophy of Science", in R. J. Seeger and R. S. Cohen, eds., *Philosophical Foundations of Science,* pp. 415–430. Dordrecht: D. Reidel.

————. 1976. "The Fertility of Theories and the Unit for Appraisal in Science", in R. S. Cohen *et al.,* eds., *Boston Studies in the Philosophy of Science* 39: 395–432. Dordrecht: D. Reidel.

Mittelstaedt, P. 1976. *Philosophical Problems of Modern Physics,* vol. 18 in *Boston Studies in the Philosophy of Science.* Dordrecht: D. Reidel.

Mittelstaedt, P., and Stachow, E. 1974. "Operational Foundations of Quantum Logic". *Foundations of Physics* 4: 355–365.

Mott, N. F., and Massey, H. S. W. 1949. *The Theory of Atomic Collisions,* 2d. ed. Oxford: Oxford University Press.

Pitowsky, I. 1982. "Substitution and Truth in Quantum Logic". *Philosophy of Science* 49: 380–401.

Putnam, H. 1977. "Realism and Reason". *Proceedings and Addresses of the American Philosophical Association* 50: 483–498.

Scheibe, E. 1973. *The Logical Analysis of Quantum Mechanics.* Oxford: Pergamon.

Sellars, W. 1979. *Naturalism and Ontology.* Reseda, Ca.: Ridgeview Publishing Co.

Shapere, D. 1982a. "Reason, Reference, and the Quest for Knowledge". *Philosophy of Science* 49: 1–23.

————. 1982b. "The Concept of Observation in Science and Philosophy". *Philosophy of Science* 49: 485–525.

Suppes, P., ed. 1980. *Studies in the Foundations of Quantum Mechanics.* East Lansing, Mich.: Philosophy of Science Association.

Thomason, R. 1970. *Symbolic Logic: An Introduction.* New York: Macmillan.

Van Aken, J. [Forthcoming.] "Elucidation of Quantum Probability Theory".

van Fraassen, B. C. 1973. "Semantic Analysis of Quantum Logic", in C. A. Hooker, ed., *Contemporary Research in the Foundations and Philosophy of Quantum Mechanics,* pp. 80–113. Dordrecht: D. Reidel.

————. 1980. *The Scientific Image.* Oxford: Clarendon Press.

Vendler, Z. 1967. *Linguistics in Philosophy.* Ithaca, N.Y.: Cornell University Press.